AMERICAN MATHEMATICAL SOCIETY
COLLOQUIUM PUBLICATIONS, VOLUME IX

DYNAMICAL SYSTEMS

BY

GEORGE D. BIRKHOFF, Ph.D., Sc.D.
PROFESSOR OF MATHEMATICS
HARVARD UNIVERSITY

NEW YORK
PUBLISHED BY THE
AMERICAN MATHEMATICAL SOCIETY
501 WEST 116TH STREET
1927

c

Lütcke & Wulff, Hamburg, Germany

upgrade

UNIVERSITY COLLEGE
UNIVERSITY
LIBRARY
NOTTINGHAM
NOTTINGHAM

12/6 G.S.

PREFACE

The Colloquium Lectures which I had the privilege of delivering at the University of Chicago before the American Mathematical Society, September 5–8, 1920, contained a large part of the material presented in the following pages. The delay in publication has been due to several causes, one of which has been my desire to wait until some of my own ideas had developed further. I have taken advantage of a well-established tradition of our Colloquia by giving particular emphasis to my own researches on dynamical systems. It is my earnest hope that the lectures may serve to stimulate others to investigate the outstanding problems in this most fascinating field.

It is only necessary to recall the work of Galileo, Newton, Laplace, Clausius, Rayleigh in the physical applications of dynamics, of Lagrange, W. R. Hamilton, Jacobi in its formal development, and of Hill and Poincaré in the qualitative treatment of dynamical questions, in order to realize the remarkable significance of dynamics in the past for scientific thought. At a time when no physical theory can properly be termed fundamental—the known theories appear to be merely more or less fundamental in certain directions—it may be asserted with confidence that ordinary differential equations in the real domain, and particularly equations of dynamical origin, will continue to hold a position of the highest importance.

In looking back over my own dynamical work, of which a certain period is finished with the publication of this book, I cannot but express my feeling of deep admiration and gratitude to Hadamard, Levi-Civita, Sundman and Whittaker, to whom many important recent advances in theoretical dynamics are due, and in whose work I have found especial inspiration. It is with much regret that I have been unable to give adequate space to their achievements.

Professor Philip Franklin coöperated with me in a first re-writing of part of my notes on these lectures. I owe him cordial thanks for his help.

November 18, 1927.

GEORGE D. BIRKHOFF.

CONTENTS

CHAPTER I

PHYSICAL ASPECTS OF DYNAMICAL SYSTEMS

CHAPTER II

VARIATIONAL PRINCIPLES AND APPLICATIONS

CHAPTER III

FORMAL ASPECTS OF DYNAMICS

CHAPTER IV

STABILITY OF PERIODIC MOTIONS

CHAPTER V

EXISTENCE OF PERIODIC MOTIONS

CHAPTER VI

APPLICATION OF POINCARÉ'S GEOMETRIC THEOREM

CHAPTER VII

GENERAL THEORY OF DYNAMICAL SYSTEMS

CHAPTER VIII

THE CASE OF TWO DEGREES OF FREEDOM

CHAPTER IX

THE PROBLEM OF THREE BODIES

CHAPTER I

PHYSICAL ASPECTS OF DYNAMICAL SYSTEMS

1. Introductory remarks. In dynamics we deal with physical systems whose state at a time t is completely specified by the values of n real variables

$$x_1, x_2, \cdots, x_n.$$

Accordingly the system is such that the rates of change of these variables, namely

$$dx_1/dt,\ dx_2/dt, \cdots, dx_n/dt,$$

merely depend upon the values of the variables themselves, so that the laws of motion can be expressed by means of n differential equations of the first order

$$(1) \qquad dx_i/dt = X_i(x_1, \cdots, x_n) \quad (i = 1, \cdots, n).$$

Thus, for a particle which falls in a vacuum at the surface of the earth, x_1 and x_2 may denote distance fallen and velocity respectively. In this case the equations of motion take the typical form

$$dx_1/dt = x_2, \qquad dx_2/dt = g,$$

where g is the gravitational acceleration.

2. An existence theorem. We proceed first to formulate an existence theorem for a set of differential equations of the general type (1).* The set of n functions X_i will be assumed to be real and uniformly continuous in some open

* In connection with the first five paragraphs the following general references may be given: E. Picard, *Traité d'Analyse*, vol. 2, chap. 11, and vol. 3, chap. 8; E. Goursat, *Cours d'Analyse mathématique*, vol. 2, chap. 19; G. A. Bliss, *Princeton Colloquium Lectures*, chap. 3.

1

finite n dimensional continuum R in the 'space' with rectangular coördinates $x_1, \cdots, x_n$. A 'solution' $x(t)$ of the equations (1) in the open interval $t' < t < t''$ is defined to be a set of n functions $x_i(t)$, all continuous together with their first derivatives and represented for any such t by a point x in R, such that the differential equations are satisfied by this set of functions.

EXISTENCE THEOREM. *If the point x^0 is in R at a distance at least D from the boundary of R, and if M is an upper bound for the functions $|X_i|$ in R, there exists a solution $x(t)$ of the equations* (1), *defined in the interval*

$$|t - t_0| < D/(\sqrt{n}\, M)$$

and for which $x(t_0) = x^0$.

To establish this theorem, we observe first that, for any solution of the type sought, the n equations

$$S_i \equiv x_i - x_i^0 - \int_{t_0}^{t} X_i(x_1, \cdots, x_n)\, dt = 0 \tag{2}$$

hold. Conversely, any set of continuous functions $x(t)$ in R, which make the expressions S_i vanish in an interval containing $t = t_0$ as an interior point, will obviously reduce to x^0 for $t = t_0$, and will satisfy the differential equations in question, as follows by direct differentiation.

Now define the set of infinitely multiple-valued functions $X_i^m(x_1, \cdots, x_n)$ as that given by *any* set $X_i(y_1, \cdots, y_n)$ taken at a point y whose various coördinates differ from those of the point x by not more than $1/m$ in numerical value. It is evident that with this definition the n components of X^m may be chosen as constant in any rectangular domain

$$|x_i - a_i| \leqq 1/m \qquad (i = 1, \cdots, n),$$

namely as the component parts of $X(a_1, \cdots, a_n)$.

If the functions X_i be replaced by X_i^m and the functions x_i by x_i^m, the expressions for S_i become

$$S_i^m \equiv x_i^m - x_i^0 - \int_{t_0}^{t} X_i^m(x_1^m, \cdots, x_n^m)\, dt.$$

We propose to show that these expressions can be made to vanish.

Choose X^m as $X(x_1^0, \cdots, x_n^0)$ in the rectangular domain

$$|x_i - x_i^0| < 1/m \qquad (i = 1, \cdots, n).$$

The integrals in the above expressions for S_i^m will then be linear functions of t, and hence x_i^m may be defined as

$$x_i^0 + X_i(x_1^0, \cdots, x_n^0)(t - t_0)$$

as long as the point x^m continues to be in this domain. In geometrical terms, the expressions for $x_i^m(t)$ yield the coördinates of a straight line with t as parameter, which passes through the center of the domain for $t = t_0$. If the n functions X_i^m happen to vanish, the line reduces to the point x^0.

In case the line emerges from the domain for $t = t_1 > t_0$ at a point y^0, we can take this point as the center of a second like rectangular domain of the same dimensions, and take x_i^m as

$$y_i^0 + X_i^m(y_1^0, \cdots, y_n^0)(t - t_1)$$

in this second domain. The expressions S_i^m will then continue to vanish for $t \geqq t_1$ until the point x^m leaves this second domain at a point z^0.

Thus, by a succession of steps, the expressions S_i^m can be made to vanish for $t > t_0$ and likewise for $t < t_0$. The process can only terminate in case the broken line representing $x^m(t)$ passes a boundary point of R.

Now, if t be taken as the time and x_i^m as the n coördinates of a particle, its velocity

$$[(X_1^m)^2 + \cdots + (X_n^m)^2]^{1/2}$$

is clearly not more than $\sqrt{n}\,M$. Hence the particle must remain inside of R at least in the interval

$$|t - t_0| < D/(\sqrt{n}\,M).$$

All of the functions x_i^m are defined in this fixed t interval whatever be the value of m.

As m takes on the values 1, 2, 3, $\cdots$, there arises an infinite sequence of sets $x_i^m(t)$ of functions defined in this interval. All of these sets lie in R, and so are uniformly bounded. Furthermore, since the S_i^m vanish for all i and m, the inequality

$$|x_i^m(t+h) - x_i^m(t)| = \left| \int_t^{t+h} X_i^m(x_1^m, \cdots, x_n^m)\, dt \right| \leqq Mh$$

obtains. Hence, by a special case of a well known theorem due to Ascoli,* there exists an infinite sequence of values of m for which every element of the set x_i^m approaches a function $\bar{x}_i$ of the set $\bar{x}$ uniformly, these functions being themselves continuous.

It is easy to prove that the functions $\bar{x}_i$ so obtained satisfy the integral form (2) of the differential equations. In fact, since the S_i^m vanish for every i and m, we have

$$\begin{aligned} \bar{S}_i &= \bar{S}_i - S_i^m \\ &= (\bar{x}_i - x_i^m) - \int_{t_0}^{t} [X_i(\bar{x}_1, \cdots, \bar{x}_n) - X_i^m(x_1^m, \cdots, x_n^m)]\, dt. \end{aligned}$$

For m sufficiently large, the first term on the right becomes uniformly small inasmuch as each $\bar{x}_i$ is approached uniformly by the corresponding x_i^m over the sequence under consideration. Also $X_i(\bar{x}_1, \cdots, \bar{x}_n)$ will differ from $X_i(x_1^m, \cdots, x_n^m)$ for any i by a uniformly small quantity, since X_i is uniformly continuous in R by hypothesis; and $X_i(x_1^m, \cdots, x_n^m)$ in turn will differ from $X_i^m(x_1^m, \cdots, x_n^m)$ by a uniformly small quantity, in virtue of the definition of the functions X_i^m. Hence the quantity under the integral sign on the right also becomes uniformly small as m increases and the expressions $\bar{S}_i$, which are independent of m, must vanish as stated, so that $\bar{x}(t)$ yields the required solution of (1).

By repeated use of the existence theorem, the given solution $x(t)$ may be extended beyond its interval of definition unless

* For a brief statement and proof see W. F. Osgood, Annals of Mathematics, vol. 14, series 2, pp. 152–153.

as t approaches either end of the interval, the corresponding point $x(t)$ approaches the boundary of R. Hence we infer the truth of the following statement:

COROLLARY. *The interval of definition for any solution $x(t)$ of the equations* (1) *may be extended so as to take one of the following four forms:*

$$-\infty < t < +\infty; \quad -\infty < t < t''; \quad t' < t < +\infty; \quad t' < t < t'',$$

where, as t approaches t' or t'', the point x approaches the boundary of R.

3. A uniqueness theorem. It may now be proved that there is only one solution of the type described in the existence theorem, in case the functions X_i possess continuous first partial derivatives. This last requirement may be lightened to a well known form given by Lipschitz.

UNIQUENESS THEOREM. *If for every i and for every pair of points x, y in R the functions X_i satisfy a Lipschitz condition,*

$$|X_i(x_1, \cdots, x_n) - X_i(y_1, \cdots, y_n)| \leqq \sum_{j=1}^{n} L_j |x_j - y_j|,$$

the quantities $L_1, \cdots, L_n$ being fixed positive quantities, then there is only one solution $x(t)$ of (1) *such that $x(t_0) = x^0$.*

For if two distinct solutions $x(t)$ and $y(t)$ have the same values x^0 for $t = t_0$, the corresponding integral forms of the differential equations give at once

$$x_i - y_i - \int_{t_0}^{t} [X_i(x_1, \cdots, x_n) - X_i(y_1, \cdots, y_n)]\, dt = 0$$

for all values of i, and thence by the Lipschitz condition imposed,

$$|x_i - y_i| \leqq \int_{t_0}^{t} \sum_{j=1}^{n} L_j |x_j - y_j|\, |dt|.$$

Let L be the maximum of the n positive constants L_i, and let Q be the maximum of any of the n quantities $|x_i - y_i|$ in any closed interval within the interval

$$|t - t_0| \leqq 1/(2nL).$$

The maximum Q must be attained for some value of t, say t^*, and for some i. If we insert the value t^* of t in the corresponding inequality above, and apply the mean value theorem to the right-hand member, there results

$$Q \leqq nLQ\,|t^* - t_0| \leqq Q/2.$$

This proves that Q must be 0. Hence the two solutions $x(t)$, $y(t)$ which coincide for $t = t_0$ will continue to do so in any such interval. The theorem follows by repeated application of this result.

The physical meaning of the existence and uniqueness theorems is evidently that the motion of a dynamical system is completely determined by the differential equations and the initial values of the variables determining the state of the system—a fact which is intuitively obvious.

Thus the treatment of a dynamical problem requires a formulation of the appropriate differential equations by means of physical principles, and a subsequent mathematical treatment of the properties of the motions on the basis of these equations.

4. Two continuity theorems. There are certain further continuity theorems which are closely allied to the two theorems established above.

First Continuity Theorem. *If the functions X_i in* (1) *satisfy a Lipschitz condition in R, the unique solution $x(t)$ for which $x(t_0) = x^0$ is a set of continuous functions of the n parameters x_i^0 and of $t - t_0$.*

We observe first that, in changing the independent variable t to $t' = t - t_0$, the modified differential equations obtained differ from (1) only in that t is replaced by t', while in the initial conditions t_0 is replaced by 0. Hence the dependent variables x_i involve t and t_0 in the combination $t - t_0$ only, so that it will suffice to prove the functions x_i to be continuous in x_i^0 and t in the case $t_0 = 0$.

A slight extension of the method used in proving the uniqueness theorem may be employed. It is apparent that if x_i and y_i are two solutions of (1) which reduce to x_i^0 and

y_i^0 respectively for $t = 0$, then by subtraction of the corresponding integral equations there is obtained

$$(x_i - y_i) = (x_i^0 - y_i^0) + \int_0^t [X_i(x_1, \cdots, x_n) - Y_i(y_1, \cdots, y_n)]\, dt,$$

provided that the value of t lies within the common interval of definition of x_i and y_i.

Suppose that x^0 lies at a distance at least D from the boundary of R, and then y^0 at a distance not more than $D/2$ from x^0. This requirement will be met if we take the maximum difference $|y_i^0 - x_i^0|$ not more than $D/(2\sqrt{n})$. Restrict t further to lie in the interval $|t| \leqq 1/(2nL)$.

Under these circumstances if Q^0 denotes the maximum difference $|x_i^0 - y_i^0|$ for any i, and Q the maximum difference $|x_i - y_i|$ in the t interval under consideration, we find for some value t^* of t by means of the above integral equations,

$$Q \leqq Q^0 + nLQ|t^*| \leqq Q^0 + Q/2.$$

Hence in the stated interval we have constantly $Q \leqq 2Q^0$, i. e., the difference $x_i - y_i$ cannot exceed twice the maximum initial difference $x_j^0 - y_j^0$ in numerical value. This means that if y^0 approaches x^0, then y approaches x uniformly throughout the stated interval. Since $|dx_i/dt| \leqq M$ everywhere, the functions x_i must be continuous in x_i^0 and t in the restricted t interval.

It remains only to remove the restriction upon the interval t.

In any closed interval of definition $0 \leqq t \leqq T$, the point $x(t)$ is throughout at a distance exceeding a positive D from the boundary of R. Consequently in a t interval of fixed length about any point t' of the selected interval, each function $x_i(t)$ and will vary continuously with $x_i(t')$ and $t - t'$. It will then be possible to select points

$$t_0 = 0,\ t_1,\ \cdots,\ t_k = T,$$

such that t_1 is in the interval about t_0, t_2 in the interval about t_1, and so on. Thus if we take $|y_i^0 - x_i^0| \leqq q$ we obtain successively

$$|x_i(t_1) - y_i(t_1)| \leqq 2\,q, \;\cdots,\; |x_i(t_k) - y_i(t_k)| \leqq 2^k\,q.$$

The truth of the theorem is now obvious for the unrestricted interval.

SECOND CONTINUITY THEOREM. *If the functions X_i admit continuous bounded first partial derivatives in R while these derivatives themselves satisfy a Lipschitz condition, then the unique solution $x(t)$ such that $x(t_0) = x^0$ has components with continuous first partial derivatives as to x_i^0 and $t - t_0$.*

To establish this theorem we resort to a consideration of the difference equality introduced at the beginning of the proof of the preceding theorem. We shall restrict y to be sufficiently near x in the interval $|t - t_0| \leqq T$, precisely as in that proof, and in addition to be such that the straight line from $x(t)$ to $y(t)$ lies in the region R for any t. The preceding theorem shows that this will be possible if $|y_i^0 - x_i^0|$ is sufficiently small.

The mean value theorem allows us then to write

$$X_i(y_1, \cdots, y_n) - X_i(x_1, \cdots, x_n) = \sum_{j=1}^{n} \frac{\partial X_i}{\partial x_j}(y_j - x_j)$$

where the arguments of $\partial X_i / \partial x_j$ are $z_{i1}, \cdots, z_{in}$ with

$$z_{ij} = x_j + \theta_i (y_j - x_j) \qquad (0 < \theta_i < 1)$$

so that z lies in R. Hence the difference equality can be written

$$y_i - x_i = y_i^0 - x_i^0 + \int_{t_0}^{t} \sum_{j=1}^{n} \frac{\partial X_i}{\partial x_j}(y_j - x_j)\,dt.$$

Suppose that $y_2^0, \cdots, y_n^0$ are taken equal to $x_2^0, \cdots, x_n^0$ respectively while y_1^0 is allowed to approach x_1^0. If we write

$$\frac{y_1 - x_1}{y_1^0 - x_1^0} = \frac{\Delta x_1}{\Delta x_1^0}, \;\cdots,\; \frac{y_n - x_n}{y_1^0 - x_1^0} = \frac{\Delta x_n}{\Delta x_1^0},$$

the n equations above gives us at once

$$\frac{\Delta x_1}{\Delta x_1^0} = 1 + \int_{t_0}^{t} \sum_{j=1}^{n} \frac{\partial X_1}{\partial x_j} \frac{\Delta x_j}{\Delta x_1^0} dt,$$

$$\frac{\Delta x_2}{\Delta x_1^0} = 0 + \int_{t_0}^{t} \sum_{j=1}^{n} \frac{\partial X_2}{\partial x_j} \frac{\Delta x_j}{\Delta x_1^0} dt,$$

$$\cdots\cdots\cdots\cdots$$

$$\frac{\Delta x_n}{\Delta x_1^0} = 0 + \int_{t_0}^{t} \sum_{j=1}^{n} \frac{\partial X_n}{\partial x_j} \frac{\Delta x_j}{\Delta x_1^0} dt.$$

For $|t - t_0|$ sufficiently small, in particular for

$$|t - t_0| \leqq 1/(2nL')$$

where L' is the upper bound of $|\partial X_i/\partial x_j|$ in R, we find that none of the integrals on the right can exceed $Q'/2$, where Q' is the maximum of any $|\Delta x_j/\Delta x_1^0|$ in this interval. By using the value of t and of the index i for the corresponding $|\Delta x_i/\Delta x_1^0|$ which yields this maximum, we establish the fact that Q' is at most 2 (compare with previous section).

Also by differentiation of the equalities above it appears that the derivatives of these ratios $\Delta x_i/\Delta x_1^0$ as to t do not exceed $2nL'$.

Consequently Ascoli's theorem can be applied to show that if Δx_1^0 be allowed to approach 0 suitably, the difference ratios will approach limits, which may be designated by y_i, and that these satisfy the integral equations

$$y_1 = 1 + \int_{t_0}^{t} \sum_{j=1}^{n} \frac{\partial X_1}{\partial x_j} y_j \, dt,$$

$$y_2 = 0 + \int_{t_0}^{t} \sum_{j=1}^{n} \frac{\partial X_2}{\partial x_j} y_j \, dt,$$

$$\cdots\cdots\cdots\cdots$$

$$y_n = 0 + \int_{t_0}^{t} \sum_{j=1}^{n} \frac{\partial X_n}{\partial x_j} y_j \, dt.$$

Evidently these conditions are equivalent to the n 'equations of variation'

$$\frac{dy_i}{dt} = \sum_{j=1}^{n} \frac{\partial X_i}{\partial x_j} y_j \qquad (i = 1, \cdots, n), \tag{3}$$

and the set of initial conditions

$$y_1(t_0) = 1, \; y_2(t_0) = 0, \cdots, y_n(t_0) = 0.$$

But these n equations and conditions, joined with the n equations (1) and attached conditions, form a system of $2n$ equations and $2n$ initial conditions in $x_1, \cdots, x_n, y_1, \cdots, y_n$, to which the existence and uniqueness theorems apply; we recall that $\partial X_i / \partial x_j$ as well as X_i satisfy Lipschitz conditions. Since the functions y_i are uniquely determined, the ratios $\Delta x_i / \Delta x_1^0$ approach the limits y_i uniformly no matter how Δx_1^0 approaches 0.*

In this way it is seen that for any i and j the partial derivatives $y_i = \partial x_i / \partial x_j^0$ exist and satisfy the equations of variation and initial conditions

$$y_1(t_0) = 0, \cdots, y_{j-1}(t_0) = 0,$$
$$y_j(t_0) = 1, \; y_{j+1}(t_0) = 0, \cdots, y_n(t_0) = 0.$$

By application of the first continuity theorem it follows that these functions $\partial x_i / \partial x_j^0$ not only exist, but are continuous in x_i^0 and $t - t_0$.

5. Some extensions. The above theorems may be extended and completed in various ways.

In the first place suppose that the functions X_i are uniformly continuous functions of $x_1, \cdots, x_n$ and a parameter c, for x in R and $c' < c < c''$, and furthermore satisfy a Lipschitz condition in the $n+1$ variables $x_1, \cdots, x_n, c$. Consider the system of $n+1$ differential equations

$$dx_i / dt = X_i(x_1, \cdots, x_n, x_{n+1}) \qquad (i = 1, \cdots, n),$$
$$dx_{n+1} / dt = 0,$$

* Otherwise, by Ascoli's theorem, another distinct set y_i could be found.

with initial conditions

$$x_i(t_0) = x_i^0 \quad (i = 1, \cdots, n), \qquad x_{n+1}(t_0) = c,$$

where x^0 is in R, and $c' < c < c''$. The existence, uniqueness and first continuity theorems apply to show that a unique solution $x_i(t)$ $(i = 1, \cdots, n+1)$ exists and is continuous in x_i^0 and c. But these functions clearly satisfy the equations and conditions

$$dx_i/dt = X_i(x_1, \cdots, x_n, c), \quad x_i(t_0) = x_0 \quad (i = 1, \cdots, n).$$

If X_i have in addition bounded first partial derivations in $x_1, \cdots, x_n, c$ satisfying Lipschitz conditions in these variables, then $\partial x_i / \partial c$ will also exist by the second continuity theorem.

Consequently the existence, uniqueness, and continuity theorems can be immediately extended to the case in which the right-hand members X_i in the differential equations (1) *contain one or more parameters.*

Again, suppose that X_i involve t as well as $x_1, \cdots, x_n$. A similar consideration of the $n+1$ differential equations

$$dx_i/dt = X_i(x_1, \cdots, x_n, x_{n+1}), \; (i = 1, \cdots, n); \; dx_{n+1}/dt = 1$$

with $n+1$ initial conditions

$$x_i(t_0) = x_i^0, \quad (i = 1, \cdots, n); \qquad x_{n+1}(t_0) = t_0$$

shows that there will exist a solution when t is suitably restricted, and that there is only one solution if the functions $X_i(x_1, \cdots, x_n, t)$ satisfy a Lipschitz condition in $x_1, \cdots, x_n, t$. Analogs of the first and second continuity theorems are easily formulated for such a system also.

Thus a similar extension is possible to the case in which the functions X_i involve the time t.

Again, let us suppose that X_i contain only the variables $x_1, \cdots, x_n$ but possess continuous partial derivatives up to those of order $\mu > 0$, while the partial derivatives of the μth order satisfy Lipschitz conditions. The method of proof of the second continuity theorem given above shows that the given system (1) of differential equations can be replaced by a system of the same type of order $2n$ with $x_1, \cdots, x_n$ and $y_1, \cdots, y_n$ where $y_i = \partial x_i / \partial x_1^0$ for instance, as dependent variables; this system of order $2n$ consists of course of the n given equations and the n equations of variation. Now if we apply the second continuity theorem to this augmented system, we conclude at once that the second partial derivatives $\partial^2 x_i / \partial x_j^0 \partial x_1^0$ and likewise $\partial^2 x_i / \partial x_j^0 \partial x_k^0$ will exist and be continuous. In the augmented system, however, the right-hand members will in general possess continuous first partial derivatives of order $\mu - 1$ which will satisfy a Lipschitz condition.

Repeating the above process we obtain the existence of partial derivatives of $x_1, \cdots, x_n$ up to those of order μ which respect to the variables $x_1^0, \cdots, x_n^0$.

In case the functions X_i of $x_1, \cdots, x_n$ possess continuous first partial derivatives of order μ while the partial derivatives of order μ satisfy Lipschitz conditions, the components $x_1, \cdots, x_n$ considered as functions of $x_1^0, \cdots, x_n^0, t - t_0$ will possess continuous partial derivatives in these variables of order μ.

An important case, and the only one entering subsequently into consideration, is that in which the functions X_i admit continuous partial derivatives of all orders in the variables concerned. The components $x_1, \cdots, x_n$ will then necessarily possess continuous partial derivatives of all orders in $x_1^0, \cdots, x_n^0, t - t_0$.

If, furthermore, the functions X_i are analytic in $x_1, \cdots, x_n$, the components $x_1, \cdots, x_n$ considered as functions of $x_1^0, \cdots, x_n^0, t - t_0$, will be analytic in these variables.

Let us indicate briefly a proof of this important fact.

We observe first that it suffices to show that the unique solution of (1) for which x reduces to x^0 for $t = 0$ has components analytic in $x_1^0, \cdots, x_n^0, t$; here the device used in the proof of the second continuity theorem, namely the introduction of $t' = t - t_0$ in the differential equations, is applicable.

Furthermore by writing

$$x_1' = x_1 - x_1^0, \cdots, x_n' = x_n - x_n^0,$$

in these equations, it becomes clear that we need only prove the components $x_1, \cdots, x_n$ to be analytic in $x_1^0, \cdots, x_n^0$ in the neighborhood of the origin.

Now since X_i are then analytic in the neighborhood of the origin, we may write

$$X_i \ll \frac{M}{1 - \dfrac{x_1 + \cdots + x_n}{r}} \qquad (i = 1, \cdots, n),$$

where M is a sufficiently large positive quantity and r is a sufficiently small positive quantity. The relations written mean that every coefficient in the series expansion of X_i in powers of $x_1, \cdots, x_n$ does not exceed the corresponding coefficient of the series on the right in numerical value.*

Now consider the comparison differential system

$$\frac{dx_i}{dt} = \frac{M}{1 - \dfrac{x_1 + \cdots + x_n}{r}} \qquad (i = 1, \cdots, n),$$

of which the unique solution which satisfies the conditions

$$x_1 = x_1^0, \cdots, x_n = x_n^0$$

for $t = 0$ is evidently given by

$$x_i = x_i^0 + u \qquad (i = 1, \cdots, n),$$

where u is defined by the implicit equation

$$\left(1 - \frac{x_1^0 + \cdots + x_n^0}{r}\right) u - \frac{nu^2}{2r} = Mt.$$

In this case $x_1, \cdots, x_n$ are clearly analytic in $x_1^0, \cdots, x_n^0, t$; furthermore the explicit formulas obtained for $x_1, \cdots, x_n$ on

* For a proof of this type of relation see, for instance, E. Picard, *Traité d'Analyse*, vol. 2, chap. 9.

successive differentiation of the comparison system and setting $x_1^0 = \cdots = x_n^0 = t = 0$ shows that the coefficients in the convergent power series for $x_1, \cdots, x_n$ in $x_1^0, \cdots, x_n^0, t$ are all positive.

But the inequality relations written above obviously imply similar relations between any partial derivatives of X_i and of the same partial derivative of the right-hand member of the comparison equations. Thus we see in succession that the formal series made out with coefficients obtained by successive differentiation of the equations (1) as to $x_1^0, \cdots, x_n^0, t$ and setting $x_1^0 \cdots = x_n^0 = t = 0$, must converge since the coefficients are less than the corresponding coefficients of known convergent series. Thus these formal series define analytic functions $x_1, \cdots, x_n$, while the mode of determination of these functions renders it certain that every difference function

$$dx_i/dt - X_i(x_1, \cdots, x_n) \qquad (i = 1, \cdots, n),$$

considered as functions of $x_1^0, \cdots, x_n^0, t$, vanishes together with all of its partial derivatives at the origin when these analytic functions are substituted in. Hence these difference functions must vanish identically. Thus $x_1, \cdots, x_n$ obtained in this formal manner will constitute the unique solution satisfying the prescribed conditions, and the stated analyticity is proved.

6. The principle of the conservation of energy.* Conservative systems. In the case of many dynamical systems the geometric configuration is determined by m 'coördinates' $q_1, \cdots, q_m$ having a spatial nature, while the state of the system is fixed by the coördinates and the velocities $q_1', \cdots, q_m'$, where $q_i' = dq_i/dt$. Such a system is said to have m 'degrees of freedom'. With these coördinates may be correlated 'generalized external forces' Q_i so that by definition

* For historical and critical remarks concerning this principle see the article by A. Voss in the *Encyklopädie der mathematischen Wissenschaften*, vol. 4 or in the French version by E. and F. Cosserat. I presented the results here obtained at the Chicago Colloquium in 1920. The following treatment of the principle differs essentially from any other which I have seen.

the 'work' W done on the system is given by

$$dW = \sum_{j=1}^{m} Q_j \, dq_j$$

in which the differential symbols have their ordinary significance.

We shall assume that the functions Q_i are real, single-valued, analytic functions of the coördinates, velocities, and accelerations; thus there is one and only one set of external forces Q_i which yields a prescribed set of accelerations for a given set of coördinate values and velocities. In this case the variables determining the state of the system are clearly the $2m$ coördinates and velocities.

As a concrete model of such a dynamical system, we may think of a concealed mechanism which is controlled by a set of m rods which project from a wall. If the rods project by distances $q_1, \cdots, q_m$, then $Q_1, \cdots, Q_m$ are the ordinary forces applied to these rods in an outward direction.

The fundamental hypothesis which embodies the principle of the conservation ot energy is that if, by any application of such external forces, the dynamical system is carried through a closed cycle, so that the set of $2m$ final values of q_i and q_i' coincides with the set of initial values, the total amount of work done on the system during the cycle vanishes. Any system of this type will be called 'conservative'.

Conservative dynamical systems can only be regarded as idealizations of the systems actually found in nature, but nevertheless they are of great importance.

Let us now consider the properties of such a conservative system. If it is carried through a cycle $ABCA$ and a modified cycle $AB'CA$, (which may be represented graphically by closed curves in the $2m$ dimensional space of the q_i and q_i'), the work done in parts ABC and $AB'C$ is the same, namely the negative of that done along the common part CA. Thus the work done along the part AC is independent of the path taken, and so depends only upon the values of $q_1, \cdots, q_m$, $q_1', \cdots, q_m'$ at C:

$$\int_A^C \sum_{j=1}^m Q_j\,d\,q_j = W(q_1, \cdots, q_m, q_1', \cdots, q_m')\Big|^C.$$

By differentiation with respect to t we obtain the following fundamental identity in the $3m$ variables q_i, q_i', q_i'':

$$\sum_{j=1}^m Q_j\,q_j' = \sum_{j=1}^m \left(\frac{\partial W}{\partial q_j} q_j' + \frac{\partial W}{\partial q_j'} q_j''\right). \tag{4}$$

This relation must subsist if the principle of the conservation of energy is to hold, and conversely it is easily seen to ensure that the principle is valid.

It is possible to give the identity an interesting explicit form. Let us endeavor to determine a function L of the $2m$ variables q_i, q_i' so that the following identity holds:

$$\sum_{j=1}^m \left[\frac{d}{dt}\left(\frac{\partial L}{\partial q_j'}\right) - \frac{\partial L}{\partial q_j}\right] q_j' = \sum_{j=1}^m \left(\frac{\partial W}{\partial q_j} q_j' + \frac{\partial W}{\partial q_j'} q_j''\right).$$

Comparing the coefficients of q_i'' on both sides, we find m conditions

$$\sum_{j=1}^m \frac{\partial^2 L}{\partial q_i' \,\partial q_j'} q_j' = \frac{\partial W}{\partial q_i'},$$

which hold if

$$\sum_{j=1}^m \left(q_j' \frac{\partial L}{\partial q_j'}\right) - L = W, \tag{5}$$

as follows by differentiation with respect to q_i'. Comparing the remaining terms which are independent of q_i'', we get the further condition

$$\sum_{i,j=1}^m \frac{\partial^2 L}{\partial q_i \,\partial q_j'} q_i' q_j' - \sum_{j=1}^m \frac{\partial L}{\partial q_j} q_j' = \sum_{j=1}^m \frac{\partial W}{\partial q_j} q_j'$$

which is obviously satisfied if L satisfies (5).

A value of L for which (5) holds can always be found. Observe first that if $Q_1, \cdots, Q_m$ can be expanded in ascending powers of the velocities $q_1', \cdots, q_m'$, then no first degree terms appear in W. That is, we have

$$W = W_0 + * + W_2 + W_3 + \cdots,$$

where the subscript indicates the degree of the term in the velocities. In fact the presence of a term W_1 would lead to a term not involving the velocities on the right-hand side of the fundamental identity whereas there is no such term on the left. If we substitute the above expansion of W and the corresponding expansion of L,

$$L = L_0 + L_1 + L_2 + \cdots$$

in the partial differential equation (5) while noting that by Euler's theorem concerning homogeneous functions,

$$\sum_{j=1}^{n} q_j' \frac{\partial V_n}{\partial q_j'} = n V_n \qquad (n = 0, 1, \cdots),$$

and if we equate terms of equal degree in the velocities, we find

$$L_0 = -W_0, \; L_2 = W_2, \; \cdots, \; L_n = \frac{W_n}{n-1}, \; \cdots,$$

while L_1 is unrestricted.

Any such function L may be called a 'principal function' associated with the arbitrary conservative system with which we started. When the linear terms in the velocities are lacking in L, a special function is obtained which has important properties.

On defining the functions R_i by means of the equations

$$(6) \qquad Q_i \equiv \frac{d}{dt}\left(\frac{\partial L}{\partial q_i'}\right) - \frac{\partial L}{\partial q_i} + R_i \qquad (i = 1, \cdots, n),$$

we observe that by the definition of L we have

$$(7) \qquad \sum_{j=1}^{m} R_j q_j' = 0.$$

Conversely, if $Q_1, \cdots, Q_m$ are of such a form that (7) obtains, the principle of the conservation of energy holds.

If W is the work function of a conservative dynamical system and if L is the associated principal function, then the generalized external forces Q_i may be written in the form (6), (7).

The italicized conclusion above may be expressed in a somewhat different fashion. As is customary, let us call a dynamical system for which $R_i = 0$ $(i = 1, \cdots, n)$ a 'Lagrangian' system. Also let us call a system $W = 0$ a 'non-energic' system. The appropriateness of the latter new term lies in the fact that whatever external forces may be applied no work can be done. In this case we may take $L = 0$ also. The alternative statement is the following:

Any conservative dynamical system has external forces which are the sum of the forces of a Lagrangian system and of a non-energic system.

Before leaving this topic we may note that for unconstrained motion we have $Q_1 = \cdots = Q_m = 0$ by definition. Here the equations of motion take the form

$$\frac{d}{dt}\left(\frac{\partial L}{\partial q_i'}\right) - \frac{\partial L}{\partial q_i} = -R_i \qquad (i = 1, \cdots, m)$$

where the quantities R_i are subject to (7). Hence we may state the following conclusion:

An unconstrained, conservative, dynamical system undergoes the same motion as a Lagrangian system to which a set of non-energic external forces is applied.

An unconstrained conservative dynamical system clearly admits an energy integral $W =$ const., which by means of (5) can be written in the alternative form

$$\sum_{j=1}^{m}\left(q_j' \frac{\partial L}{\partial q_j'}\right) - L = \text{const.}$$

Lagrangian and non-energic systems have been defined by means of the types of the external forces. These definitions are not mutually exclusive. In fact let us inquire when a dynamical system will be both non-energic and Lagrangian. Since it is non-energic, we have

$$W = W_0 + W_2 + \cdots = 0,$$

and hence we find the most general function

$$L = L_1 = \sum_{j=1}^{m} \alpha_j q_j'.$$

Since the system is Lagrangian, we may take $R_i = 0$ for every i, and thus find directly

$$Q_i = \sum_{j=1}^{m} \left(\frac{\partial \alpha_i}{\partial q_j} - \frac{\partial \alpha_j}{\partial q_i} \right) q_j' \quad (i = 1, \cdots, n).$$

Hence for a system to be of both types the generalized external forces must have this specific form.

It is worthy of note that in the case $m = 1$, equation (7) implies that R_1 is zero, so that any conservative dynamical system with a single degree of freedom is Lagrangian. Similarly in the case $m = 2$ the external forces may be represented in the form

$$Q_1 = \frac{d}{dt}\left(\frac{\partial L}{\partial q_1'}\right) - \frac{\partial L}{\partial q_1} + \lambda q_2', \quad Q_2 = \frac{d}{dt}\left(\frac{\partial L}{\partial q_2'}\right) - \frac{\partial L}{\partial q_2} - \lambda q_1',$$

where λ is an arbitrary function of the coördinates, velocities, and accelerations.

7. Change of variables in conservative systems. In the first instance the coördinates q_i of a conservative dynamical system are actual distances, while the Q_i are forces which act in the direction of these coördinates. But for most physical purposes it is not desirable to adhere to a single set of coördinates.

Let us now *define* the modified external forces $\overline{Q}_i$ corresponding to the new coördinates $\overline{q}_i$ by means of the equations

$$\overline{Q}_i = \sum_{j=1}^{n} Q_j \frac{\partial q_j}{\partial \overline{q}_i} \qquad (i = 1, \cdots, n). \tag{8}$$

When this definition is adopted and a further change of variables from $\overline{q}_i$ is made, while the new external forces are defined in terms of $\overline{Q}_i$ by analogous formulas, it is found that the final functions Q_i obtained have the same expressions as if only a single direct change of variables is made. This group property is an immediate consequence of the above definition.

Hence the $\overline{Q}_i$ are uniquely defined for any coördinate system whatsoever.

It may be observed that for change from one rectangular system to another the above formula for the determination of the $\overline{Q}_i$ in terms of Q_i agrees with that obtained by the ordinary laws for the composition of forces. In more general cases the equations above define the generalized force components in the appropriate sense.

Now on account of the identity

$$dW = \sum_{j=1}^{m} Q_j\, dq_j = \sum_{j=1}^{m} \overline{Q}_j\, d\overline{q}_j,$$

it follows that the system will remain a conservative dynamical system according to our definition, in these new coördinates. Furthermore the modified work function will be the same as before (up to an additive constant), and since the formulas of transformation of the velocities

$$q_i' = \sum_{j=1}^{m} \frac{\partial q_i}{\partial \overline{q}_j}\, \overline{q}_j' \qquad (i = 1, \cdots, m)$$

are linear and homogêneous in the velocities, it follows that the various components $W_0, W_2, \cdots$ in W will be unaltered.

If we agree for the sake of definiteness always to choose the unique determination of L which lacks linear terms in the velocities, it follows that the principal function L is the same in both problems.

Suppose now that we define $\overline{R}_i$ by means of the equations

$$\overline{Q}_i = \frac{d}{dt}\left(\frac{\partial L}{\partial \overline{q}_i'}\right) - \frac{\partial L}{\partial \overline{q}_i} + \overline{R}_i$$

where L is the first given principal function, but expressed in terms of the new variables $\overline{q}_i, \overline{q}_i'$.

It is easy to prove the formal identities

$$\sum_{j=1}^{m}\left[\frac{d}{dt}\left(\frac{\partial \varphi}{\partial q_j'}\right)-\frac{\partial \varphi}{\partial q_j}\right]\frac{\partial q_j}{\partial \bar{q}_i}=\frac{d}{dt}\left(\frac{\partial \varphi}{\partial \bar{q}_i'}\right)-\frac{\partial \varphi}{\partial \bar{q}_i} \qquad (i=1,\cdots,n),$$

where on the left side φ is any function of q_i, q_i'. To see this, we note that from the linear relationship written between the variables q_i' and $\bar{q}_i'$ we have

$$\frac{\partial q_i'}{\partial \bar{q}_j'}=\frac{\partial q_i}{\partial \bar{q}_j}, \quad \frac{\partial q_i'}{\partial \bar{q}_j}=\frac{d}{dt}\left(\frac{\partial q_i}{\partial \bar{q}_j}\right) \quad (i,j=1,\cdots,n).$$

Hence we deduce for any i

$$\begin{aligned}\sum_{j=1}^{m}\left[\frac{d}{dt}\left(\frac{\partial \varphi}{\partial q_j'}\right)\right]\frac{\partial q_j}{\partial \bar{q}_i} &= \sum_{j=1}^{m}\left[\frac{d}{dt}\left(\frac{\partial \varphi}{\partial q_j'}\frac{\partial q_j'}{\partial \bar{q}_i'}\right)-\frac{\partial \varphi}{\partial q_j'}\frac{d}{dt}\left(\frac{\partial q_j}{\partial \bar{q}_i}\right)\right]\\ &= \frac{d}{dt}\left(\frac{\partial \varphi}{\partial \bar{q}_i'}\right)-\sum_{j=1}^{m}\frac{\partial \varphi}{\partial q_j'}\frac{d}{dt}\left(\frac{\partial q_j}{\partial \bar{q}_i}\right).\end{aligned}$$

Moreover we have also for any i

$$\sum_{j=1}^{m}\frac{\partial \varphi}{\partial q_j}\frac{\partial q_j}{\partial \bar{q}_i}=\frac{\partial \varphi}{\partial \bar{q}_i}-\sum_{j=1}^{m}\frac{\partial \varphi}{\partial q_j'}\frac{\partial q_j'}{\partial \bar{q}_i}.$$

Subtracting the two identities thus obtained we obtain the specified identity.

This identity with $\varphi = L$ shows of course that the functions $\bar{R}_i$ as obtained from R_i by the defining formulas have the same structure as those which give $\bar{Q}_i$ in terms of Q_i. Hence we are led to the following general result:

If the variables $q_1, \cdots, q_m$ of a conservative dynamical system are transformed to $\bar{q}_1, \cdots, \bar{q}_m$, the system remains conservative in the new variables with L, W unaltered, while Q_i, R_i are both modified to corresponding new expressions obtained as in (8). *In particular then if the system is Lagrangian or non-energic in the first set of variables, it remains so in the modified variables.*

8. Geometrical constraints. We are now in a position to deal with the question of 'geometrical constraints.' Let us suppose that various geometrical points of the given conservative system are fixed or are constrained to lie in smooth curves or surfaces or to move subject to connections by various rigid bars without mass.

The effect of such constraints is to reduce the number of degrees of freedom. In fact, by properly taking coördinates $q_1, \cdots, q_m$, the k conditions of constraint may be made to take the form

$$q_{\mu+1} = \text{const.}, \cdots, q_m = \text{const.} \qquad (\mu = m - k).$$

Now denote by $\bar{L}$ that which L becomes when the k constrained coördinates $q_{\mu+1}, \cdots, q_m$ have these assigned constant values, while the corresponding q_i', q_i'' vanish of course. Then it is clear that

$$L = \bar{L}, \quad \frac{\partial \bar{L}}{\partial q_i} = \frac{\partial L}{\partial q_i}, \quad \frac{\partial \bar{L}}{\partial q_i'} = \frac{\partial L}{\partial q_i'} \qquad (i = 1, \cdots, \mu).$$

Hence we have the relations

$$Q_i = \frac{d}{dt}\left(\frac{\partial \bar{L}}{\partial q_i'}\right) - \frac{\partial \bar{L}}{\partial q_i} + R_i \qquad (i = 1, \cdots, \mu),$$

where Q_i and R_i are defined as usual.

But the original external forces Q_i may be decomposed into a sum

$$\bar{Q}_i + P_i, \qquad (i = 1, \cdots, m),$$

in which the 'forces of constraint' P_i can do no work for any possible displacement subject to the constraints. It follows that the functions $P_1, \cdots, P_\mu$ must vanish when the coördinates are selected as above. Hence we may replace Q_i by $\bar{Q}_i$ in the formula above for $i = 1, \cdots, \mu$.

Thus we are led to the following conclusion:

If a conservative dynamical system with m degrees of freedom is subject to k geometrical constraints, it may be treated as such a conservative system with $m - k$ degrees of freedom.

9. Internal characterization of Lagrangian systems. In most of the dynamical applications, Lagrangian systems can be regarded as dealing with a system of particles subject to certain forces and geometrical constraints. This type of internal characterization formed the basis of Lagrange's derivation of these equations, and is considered briefly in the present section. In the next section an external characterization is developed.

We shall begin by considering three particular types of particles in ordinary space:

(a) *The inertial particle.*

Here if x, y, z are the rectangular coördinates of the particle, the external forces X, Y, Z in the directions of the corresponding axes are proportional to the accelerations in these directions:

$$X = mx'', \qquad Y = my'', \qquad Z = mz'',$$

where the constant of proportionality, m, is termed the 'mass' of the particle.

This is the case of an ordinary mass particle.

The particle is seen to be of Lagrangian type with Lagrangian function

$$L = \frac{1}{2} m (x'^2 + y'^2 + z'^2),$$

and L is its 'kinetic' energy.

(b) *The non-kinetic particle.*

Such a particle is subject to forces independent of the velocity and having the particular form:

$$X = -\partial V/\partial x, \qquad Y = -\partial V/\partial y, \qquad Z = -\partial V/\partial z$$

where V depends on the coördinates of the particle in space. Here the dynamical system is Lagrangian with $L = V$.

The function $-V$ is the 'potential' energy of the particle due to the field of force in which the particle moves.

An electrified particle of slight mass moving in a static electric field is nearly of this type.

(c) *The gyroscopic particle.*

By definition the gyroscopic particle is subject to forces with components of the type

$$X = (\partial \alpha/\partial y - \partial \beta/\partial x)y' + (\partial \alpha/\partial z - \partial \gamma/\partial x)z',$$

so that the force vector is perpendicular to the velocity vector and therefore can do no work. Nevertheless the system is Lagrangian with

$$L = \alpha x' + \beta y' + \gamma z'.$$

It will be noted that this is the type of special Lagrangian system which is also non-energic.

An electrically charged particle of negligible mass moving in a static magnetic field falls under this case.

(d) *The system of generalized particles.*

If a particle moves subject to a sum of forces of the inertial, non-kinetic, and gyroscopic types it may be termed a generalized particle. Such a situation is realized for example when an ordinary mass particle moves in a gravitational field. It is evident that the resultant system will then be Lagrangian with a principal function merely the sum of the principal functions associated with the component forces.

Consider further a set of such particles which do not at first interact in any way. If we add together the Lagrangian functions for the several particles, there is obtained a single function L which can serve as a single function from which the equations of motion of the system of particles may be derived.

It is necessary of course to use suitable variables (x_i, y_i, z_i) where $i = 1, 2, \cdots, m$ to differentiate between the coördinates of the various particles.

Clearly this yields a principal function L which will be quadratic in the velocities. It is another step in the way of generalization to take L to be any quadratic polynomial in the velocities, in which the homogeneous quadratic part is the kinetic energy T, in which the term U independent of the velocities is the potential energy, and in which the homogeneous term of first degree may be called the 'gyroscopic energy'.

Furthermore, as has been seen above, we may suppose the particles to be subject to certain types of geometrical constraints, thus diminishing the number of degrees of freedom, without affecting the Lagrangian character of the problem.

It is such a system of generalized particles which suffices for most of the applications.

(e) *The generalized particle in m-dimensional space.*

By an obvious extension into which we shall not enter here it appears that a single mass particle lying on an m-dimensional manifold defined by a quadratic differential form, subject to a field of force derived from a potential function in the surface, and furthermore to gyroscopic forces derived from some linear function of the velocities in the surface, will be of Lagrangian type. The function L is quadratic in the velocities. Conversely any Lagrangian system with m degrees of freedom for which L is quadratic in the velocities is representable by the motion of a mass particle in such an m-dimensional manifold.

Thus we may interpret the motion of any dynamical system with m degrees of freedom as isomorphic with the motion of a single generalized particle on a suitable m-dimensional surface.

10. External characterization of Lagrangian systems.* In this section we propose to characterize an important type of Lagrangian systems by means of certain simple properties ot the external forces.

In fact, we shall characterize those 'regular' dynamical systems for which the Lagrangian function L is a quadratic function of the velocities, without first degree terms. These form an important class of dynamical systems in which L has the form $T - U$ where T is homogeneous and quadratic

* The material of this section was presented before the Chicago Colloquium in 1920. For an analytic characterization of the Lagrangian system in the case when the external forces are linear in the velocities see E. T. Whittaker, *Analytical Dynamics*, p. 45.

in the velocities, while U is a function of the coördinates only. It will be observed that regular systems remain of this type under an arbitrary transformation of the coördinates $q_1, \cdots, q_m$.

The first of the characteristic properties which we shall state is the following:

I. *The external forces vary linearly with the coördinate accelerations.*

Evidently this means that we may write.

$$Q_i = \sum_{j=1}^{m} a_{ij} q_j'' + b_i$$

where a_{ij}, b_i do not involve the accelerations.

II. Principle of Reciprocity. *The change in the acceleration q_j'' due to a change in the i-th force Q_i is the same as the change in the acceleration q_i'' due to an equal change in the j-th force Q_j, $(i, j = 1, \cdots, n)$.**

In order to see what this means we suppose that Q_k receives a certain acceleration increment Q, in which case the above equations give

$$Q \delta_{ik} = \sum_{j=1}^{m} a_{ij} \Delta_1 q_j'' \qquad (i = 1, \cdots, m)$$

where Δ denotes the increment as usual and where $\delta_{ik} = 1$ for $i = k$ and $\delta_{ik} = 0$ for $i \neq k$.

Now suppose that Q_l receives the same increment. We find similarly

$$Q \delta_{il} = \sum_{j=1}^{m} a_{ij} \Delta_2 q_j''.$$

If we assume that the determinant $|a_{ij}|$ is not zero, we may solve these equations and obtain for all i, k, l

$$\Delta_1 q_i' = \sum_{j=1}^{m} \bar{a}_{ij} Q \delta_{jk} = \bar{a}_{ik} Q, \; \Delta_2 q_i' = \bar{a}_{il} Q,$$

where $\bar{a}_{ij}$ is the cofactor of the element in the j-th row and i-th column of $|a_{ij}|$, divided by this determinant. Putting

* Compare Rayleigh, *Theory of Sound*, vol. 1, chap. 4.

$i = l$ and $i = k$ respectively, we obtain from II, $\bar{a}_{lk} = \bar{a}_{kl}$. Thus the cofactors $\bar{a}_{ij}$ are symmetric in i and j, whence it follows that the elements a_{ij} are symmetric also, i. e. we must have $a_{ij} = a_{ji}$ for all values of i and j.

III. *For a family of similar motions, the forces are quadratic functions of the speed.*

In other words, let $q_i = q_i(t)$, $(i = 1, \cdots, m)$ be a motion of the system. Suppose that the motion is speeded up in the ratio λ to 1. The external forces become

$$Q_i = \sum_{j=1}^{m} a_{ij}(q_1, \cdots, q_m, \lambda q_1', \cdots, \lambda q_m')\lambda^2 q_j'' + b_i(q_1, \cdots, q_m, \lambda q_1', \cdots, \lambda q_m'),$$

inasmuch as the coördinates q_i are unaltered while the velocities q_i' and the accelerations q_i'' are multiplied by λ and λ^2 respectively.

If these expressions Q_i are to be quadratic in λ (q_i, q_i', q_i'' being entirely independent variables of course), the functions a_{ij} cannot depend on the velocities, while b_i will be quadratic in them. Thus in virtue of III we obtain more precisely

$$Q_i = \sum_{j=1}^{m} a_{ij}\, q_j'' + \sum_{j,k=1}^{m} b_{ijk}\, q_j' q_k' + \sum_{j=1}^{m} b_{ij}\, q_j' + b_i$$

where the functions a_{ij}, b_{ijk}, b_{ij}, b_i depend only upon the coördinates $q_1, \cdots, q_m$.

This form is rendered still more specific by the following hypothesis.

IV. Reversibility. *Any motion under prescribed external forces may equally be described in the reversed order of time.*

The meaning here is that the above relations continue to hold when t is replaced by $-t$. But this changes the velocities to their negatives while leaving the coördinates q_i and the accelerations q_i'' unaltered. We infer that the functions b_{ij} must be lacking. Hence we may write

$$Q_i = \sum_{j=1}^{m} a_{ij}\, q_j'' + \sum_{j,\,k=1}^{m} b_{ijk}\, q_j'\, q_k' + b_i$$

where a_{ij}, $b_{ijk} = b_{ikj}$, b_i involve only the coördinates.

All of the properties I to IV so far employed are invariant under a change of coördinates q_i and have to do with the nature of the external forces in the neighborhood of a set of values $q_1^0, \cdots, q_m^0$.

By a suitable choice of coördinates at a point $q_1^0, \cdots, q_m^0$, we can reduce the expressions for $Q_1, \cdots, Q_m$ to the simple form

$$Q_i = q_i'' + b_i \qquad (i = 1, \cdots, m)$$

at that point.

To establish this fact we assume that $q_1^0, \cdots, q_m^0$ is at the origin, and make a first linear transformation

$$q_i = \sum_{j=1}^{m} \beta_{ij}\, \bar{q}_j \qquad (i = 1, \cdots, m),$$

where the β_{ij} are constants with $|\beta_{ij}| \neq 0$. For the functions $\bar{Q}_i$ we have then

$$\bar{Q}_i = \sum_{j=1}^{m} Q_j \frac{\partial q_j}{\partial \bar{q}_i} = \sum_{j=1}^{m} Q_j\, \beta_{ji}.$$

Hence we find by substitution for every i,

$$\bar{Q}_i = \sum_{j,\,k,\,l=1}^{m} a_{jk}\, \beta_{ji}\, \beta_{kl}\, \bar{q}_l'' + \text{terms independent of } q_1'', \cdots, q_m''.$$

It follows that if we choose our transformation so as to transform the quadratic form

$$\sum_{j,\,k=1}^{m} a_{jk}\, q_j\, q_k$$

into the sum of squares

$$q_1^2 + \cdots + q_m^2,$$

the quantities $\bar{a}_{ij}$ will be 0 for $i \neq j$ and 1 for $i = j$. Consequently it is at least legitimate to assume that a_{ij} has been reduced to δ_{ij} at the origin by this preliminary transformation, so that we have

$$Q_i = q_i'' + \sum_{j,k=1}^{m} b_{ijk}^0 q_j' q_k' + b_i^0 \qquad (i = 1, \cdots, m)$$

at the origin.

Now suppose that we write further

$$\bar{q}_i = q_i + \frac{1}{2} \sum_{j,k=1}^{m} b_{ijk}^0 q_j q_k$$

where the constants b_{ijk}^0 have the values specified in the equations above. We find by differentiation that the equations

$$\bar{q}_i' = q_i', \quad \bar{q}_i'' = q_i'' + \sum_{j,k=1}^{m} b_{ijk}^0 q_j' q_k' \qquad (i = 1, \cdots, m)$$

hold at the origin.

It is then found at once that in these variables $\bar{q}_i$, the formula for $\bar{Q}_i$ has the stated form at the origin.

V. Conservation of Energy. *The dynamical system is conservative.*

If W is the work function we have the fundamental relation

$$dW = \sum_{j=1}^{m} Q_j q_j' dt,$$

characteristic of conservative systems. But the sum on the right is linear in the accelerations, and comparing the coefficients of q_j'' obtained by employing the form for Q_i above we find

$$\partial W / \partial q_j' = \sum_{i=1}^{m} a_{ij} q_i' \qquad (j = 1, \cdots, m),$$

whence

$$W = T + U$$

where

$$T = \frac{1}{2} \sum_{j,k=1}^{m} a_{jk} q_j' q_k',$$

and where U is a function of $q_1, \cdots, q_m$ only.

Employing this more specific form for W, we have of course $L = T - U$ according to the earlier developments, and thence

$$Q_i = \frac{d}{dt}\left(\frac{\partial T}{\partial q_i'}\right) + \frac{\partial U}{\partial q_i} + R_i \qquad (i = 1, \cdots, m)$$

where R_i satisfy (7). But the first two terms on the right yield expressions just like those for those derived above for Q_i, the terms in q_j'' being identical. It follows that we must have the differences R_i of the form

$$R_i = \sum_{j,k=1}^{m} c_{ijk}\, q_j' q_k' + c_i \qquad (i = 1, \cdots, m).$$

Applying now the above condition (7) we conclude further that for all i, j, k the relations

$$c_{ijk} + c_{jki} + c_{kij} = 0, \qquad c_i = 0$$

must obtain where $c_{ijk} = c_{ikj}$ also of course.

Hence principles I–V lead to the type of external forces,

$$Q_i = \frac{d}{dt}\left(\frac{\partial T}{\partial q_i'}\right) + \frac{\partial U}{\partial q_i} + \sum_{j,k=1}^{m} c_{ijk}\, q_j' q_k',$$

where the $c_{ijk} = c_{ikj}$ are functions of the coördinates such that for all i, j, k

$$c_{ijk} + c_{jki} + c_{kij} = 0.$$

It remains to specify a final condition, as simple as possible, which will allow us to conclude $c_{ijk} = 0$ for all i, j, k.

VI. *If by a particular choice of coördinates, the kinetic energy T is made stationary in $q_1, \cdots, q_m$ at a certain point $q_1^0, \cdots, q_m^0$, then the forces Q_i yield accelerations which are independent of the velocities.*

Suppose for a moment that such a stationary T exists, so that we have at $q_1^0, \cdots, q_m^0$

$$\partial a_{ij} / \partial q_k = 0 \qquad (i, j, k = 1, \cdots, m).$$

The form for Q_i at this point becomes

$$Q_i = \sum_{j=1}^{m} a_{ij}\, q_j'' + \frac{\partial U}{\partial q_i} + \sum_{j,k=1}^{m} c_{ijk}\, q_j'\, q_k'.$$

It is to be observed that the form of Q_i employed holds for any coördinate system. Now if these forces Q_i are to be independent of the velocities, we must have $c_{ijk} = 0$ for all i, j, k in the special coördinate system and thus in the most general system by the known law of transformation of the terms R_i. Consequently the desired Lagrangian form of external forces is obtained.

The hypothesis that a stationary T exists is justified by the well-known fact that for any coördinates of geodesic type at $q_1^0, \cdots, q_m^0$, the surface element ds where

$$ds^2 = \sum_{i,j=1}^{m} a_{ij}\, d\,q_i\, d\,q_j$$

has coefficients a_{ij} stationary at this point.

Conversely, it is readily seen that a regular Lagrangian system has external forces Q_i which satisfy I–VI.

11. Dissipative systems. Conservative systems are often limiting cases of what is found in nature, since actual work is usually done on the system during a closed cycle. A system for which work is done may be called dissipative. More explicitly we shall define dissipative systems to be such that

$$Q_i = \frac{d}{dt}\left(\frac{\partial L}{\partial q_i'}\right) - \frac{\partial L}{\partial q_i} + R_i \quad (i = 1, \cdots, m)$$

where

$$\sum_{j=1}^{m} R_j\, q_j' \geqq 0.$$

Furthermore we shall assume that the equality sign can only hold for motions in a manifold of dimensionality less than m in the m-dimensional coördinate space.

Suppose now that such a system is unconstrained, or at least is subject to external forces which do no work, so that

$$\sum_{j=1}^{m} Q_j\, q_j' = 0.$$

Because of the obvious relation

$$\frac{dW}{dt} + \sum_{j=1}^{m} R_j q'_j = 0,$$

where W denotes the work function associated with L, namely

$$\sum_{j=1}^{m} \left(q'_j \frac{\partial L}{\partial q'_j} \right) - L,$$

we infer that W constantly diminishes toward some limiting value W_0. It is assumed that the work function cannot diminish to $-\infty$.

Now consider the limiting motions of the given motion. Along these motions W has this limiting value W_0, and of course the sum

$$\sum_{j=1}^{m} R_j q'_j$$

vanishes.

A dissipative system of this type tends in its unconstrained motion either toward equilibrium or, more generally, toward the motion of a conservative system with fewer degrees of freedom.

CHAPTER II

VARIATIONAL PRINCIPLES AND APPLICATIONS

1. An algebraic variational principle. On the formal side of dynamics it has proved to be a fact of fundamental importance that the differential equations can in general be obtained by demanding that the 'variation' of some definite integral vanishes.

To make clear the essential nature of the variational method, we may consider an analogous question concerning ordinary maxima and minima.

Let there be given n equations in n unknown quantities.

$$f_i(x_1, \cdots, x_n) = 0 \qquad (i = 1, \cdots, n),$$

in which the left hand-members are expressible as the partial derivatives of a single unknown real analytic function F,

$$f_i = \partial F / \partial x_i \qquad (i = 1, \cdots, n)$$

The n equations are then of the special type which arises in the determination of the maxima and minima of F, and they may be combined in one symbolic equation $dF = 0$. Their significance is that for the values $x_1^0, \cdots, x_n^0$ under consideration, the function F is 'stationary'.

Now suppose that the variables x_i are changed to y_i in the n equations, where the relation between x_i and y_i is one-to-one and analytic. Since the phenomenon of a stationary value of F is clearly independent of the particular variables in terms of which F is expressed, the solutions of the original equations can be expressed in the characteristic differential form $dF = 0$, in the new as well as the old variables. This furnishes a means of obtaining an equivalent system of equations in the new variables, which is in general simpler than that of direct substitution in the original equations.

In cases when it is not possible to write the given equations in the special form, it is frequently possible to find combinations of these equations which may be so written.

Moreover any non-specialized set of n equations in $x_1, \cdots, x_n$ of the form first written is equivalent to $2n$ equations obtained from $dF = 0$ where

$$F = \sum_{j=1}^{n} f_j x_{n+j},$$

at least provided that the determinant $|\partial f_i / \partial x_j| \neq 0$. For we find that $x_{n+1}, \cdots, x_{2n}$ are 0, while $x_1, \cdots, x_n$ must satisfy the required equations.

From these circumstances it is easy to conjecture that the significance of the analogous variational principles of dynamics is largely formal.

2. Hamilton's principle. Let us formulate the concept of a 'stationary integral'. Suppose that the equations

$$x_i = x_i(t, \lambda) \qquad (i = 1, \cdots, m)$$

represent a family of functions depending on the parameter λ in such wise that for $\lambda = 0$ we have a given set of functions,

$$x_i(t, 0) = x_i^0(t) \qquad (i = 1, \cdots, m).$$

We shall assume that the functions $x_i(t, \lambda)$ are continuous with continuous first and second partial derivatives in t and λ, and also that these functions of t and λ vanish identically sufficiently near to the two ends of the interval (t_0, t_1) under consideration,

$$x_i(t, \lambda) = 0 \qquad (t_0 \leqq t \leqq t_0 + \varepsilon,\ t_1 - \varepsilon \leqq t \leqq t_1).$$

Under these conditions the integral

$$I = \int_{t_0}^{t_1} F(x_1, \cdots, x_m, x_1', \cdots, x_m')\, dt,$$

where F and its partial derivatives of the first two orders are taken continuous, is said to be 'stationary' for $x_i = x_i^0(t)$, if for every such family of functions we have

$$\delta I = \left. \frac{\partial I}{\partial \lambda} \right|_{\lambda=0} \delta \lambda = 0.$$

This amounts to the equation for $\lambda = 0$,

$$\int_{t_0}^{t_1} \sum_{j=1}^{m} \left(\frac{\partial F}{\partial x_j} \frac{\partial x_j}{\partial \lambda} + \frac{\partial F}{\partial x_j'} \frac{\partial x_j'}{\partial \lambda} \right) dt = 0.$$

Integrating by parts and noting that δx_i vanishes at the end points, we obtain the equivalent equations

$$\int_{t_0}^{t_1} \sum_{j=1}^{m} \left[\frac{\partial F}{\partial x_j} - \frac{d}{dt} \left(\frac{\partial F}{\partial x_j'} \right) \right] \delta x_j \, dt = 0.$$

In particular we may take

$$x_i(t, \lambda) = x_i^0(t) + \lambda \delta x_i \quad (i = 1, \cdots, m)$$

where the functions δx_i are arbitrary continuous functions of t with a continuous first and second derivative except that they are to vanish near to t_0 and t_1.

In this way the condition that the integral be stationary is found to be equivalent to the system of m differential equations of Euler in $x_1^0, \cdots, x_m^0$,

$$\frac{d}{dt} \left(\frac{\partial F}{\partial x_i'} \right) - \frac{\partial F}{\partial x_i} = 0 \quad (i = 1, \cdots, m).$$

In fact the above integral can not vanish for all possible admissible functions $x_i(t, \lambda)$ unless this condition is satisfied.*

But the m equations just written are identical in form with the Lagrangian equations except that L is replaced by F. Hence we obtain the following important result:

* See, for example, O. Bolza, *Vorlesungen über Variationsrechnung*, chap. 1, for fuller statements and arguments.

The Lagrangian equations may be given the variational form known as Hamilton's principle,

$$\delta \int_{t_0}^{t_1} L\, dt = 0. \tag{1}$$

According to the principle which led us to introduce the concept of variation, we may affect any desired change of variables in the given Lagrangian equations by introducing the new variables in the function L. To this fact is due much of the convenience of the Lagrangian form.

3. The principle of least action. There is a second well-known variational form for the Lagrangian equations termed the 'principle of least action', and we proceed to clarify the relation of this principle to the one just formulated. We assume that $L = L_2 + L_1 + L_0$ is quadratic in the velocities, and recall that the Lagrangian equations admit the energy integral

$$W \equiv \sum_{j=1}^{m} \left(q_j' \frac{\partial L}{\partial q_j'} \right) - L = L_2 - L_0 = c.$$

It is on this fact that our considerations will be based.

Let us confine attention to the case where the energy constant c has a specified value, say $c = 0$. Hence we have $L_2 = L_0$ along the motion $q_i = q_i^0(t)$, $(i = 1, \cdots, m)$, considered.

Now define I^* as follows:

$$I^* = I - \int_{t_0}^{t_1} (\sqrt{L_2} - \sqrt{L_0})^2\, dt = \int_{t_0}^{t_1} (2\sqrt{L_0 L_2} + L_1)\, dt.$$

This yields

$$\delta I^* = \delta I - 2\int_{t_0}^{t_1} (\sqrt{L_2} - \sqrt{L_0})\,(\delta\sqrt{L_2} - \delta\sqrt{L_0})\, dt.$$

Accordingly, if the $q_i^0(t)$ satisfy the assumed energy condition we shall have

$$\delta I^* = \delta I$$

for *all* variations of the q_i. Hence if the q_i^0 in addition satisfy the Lagrangian equations, so that $\delta I = 0$, we shall have $\delta I^* = 0$ also.

The integrand of I^* is positively homogeneous of dimensions unity in the derivatives q_i'. Consequently the numerical value of this integral I^* is independent of the parameter t used along the path of integration, and the value of the integral depends only on the path in $q_1, \cdots, q_m$ space;† for variations of the admitted type the end points of the path are fixed. Thus the integral of energy can be regarded as merely determining the parameter t, since if we write

$$\bar{t} = \int^{t} \sqrt{L_2}/\sqrt{L_0}\, dt,$$

the integral relation is satisfied in the new parameter $\bar{t}$.

Consequently, if we have $\delta I^* = 0$ for $q_i = q_i^0(t)$ and if the new parameter t is chosen in this manner, we have $\delta I = 0$ also for $q_i = q_i^0(t)$.

An alternative variational form for the equations of motion of such a Lagrangian system is $\delta I^* = 0$, *or more explicitly,*

$$(2) \qquad \delta \int_{t_0}^{t_1} (2\sqrt{L_0 L_2} + L_1)\, dt = 0$$

provided that L_0 *is so chosen that the energy constant vanishes, and the parameter* t *is determined as specified.*

The equation $\delta I^* = 0$ constitutes the 'principle of least action' for this problem, and is usually given for the case where the linear term L_1 in the velocities is not present.

By means of this principle not only the variables q_i but also the variable t may be transformed with facility. Indeed, it is obvious that the condition $\delta I^* = 0$ is invariant in form under a transformation of the dependent variables q_i to new variables $\bar{q}_i$. For along the transformed curve the same variational condition will be satisfied, except that L is replaced by its expression in terms of the new variables, while t has the same meaning as before. Consequently in order to transform these variables, it is sufficient to effect the transformation of L directly. The corresponding trans-

† See O. Bolza, *Vorlesungen über Variationsrechnung*, chap. 5.

formed equations are then obtained by the use of the new expression for L.

The allowable type of transformation of the independent variable t is the following:

$$dt = \mu(q_1, \cdots, q_m)\, d\bar{t}.$$

In other words, the differential element of time is divided by a factor μ depending upon the coördinates. We may determine the nature of the modification which the Langrangian equations undergo as a result of this transformation as follows: We note that the integral I^* may be written equally well

$$I^* = \int_{\bar{t}_0}^{\bar{t}_1} (2\sqrt{\mu L_0 \cdot \mu L_2} + \mu L_1)\, d\bar{t}.$$

This modified integral is of the same form as before if we set

$$\bar{L} = \mu L.$$

Furthermore δI^* vanishes along the curve whether t or $\bar{t}$ be regarded as parameter. By this transformation of t, then, the equations of Lagrange and the given integral condition go over into other equations of the same type with the principal function L multiplied by μ.

The differential form $L\,dt$ is invariant under transformations of either type. We conclude therefore the following fact: *By a transformation*

$$q_i = f_i(\bar{q}_1, \cdots, \bar{q}_m) \quad (i = 1, \cdots, m), \qquad dt = \mu(\bar{q}_1, \cdots, \bar{q}_m)\, d\bar{t},$$

the Lagrangian equations with energy constant 0 *go over into a like set of equations with energy constant* 0 *in which* $\bar{L}$ *is obtained from the formula*

$$L\,dt = \bar{L}\,d\bar{t}.$$

In the reversible case we have $L_1 = 0$, and thus

$$I^* = \int_{t_0}^{t_1} 2\sqrt{L_0 L_2}\, dt = 2\int_{t_0}^{t_1} ds,$$

where $ds^2 = L_0 L_2 (dt)^2$ is the squared element of arc on a surface with coördinates $q_1, \cdots, q_m$.

Thus in the reversible case with fixed energy constant the curves of motion may be interpreted as geodesics on the m-dimensional surface with squared element of arc

$$ds^2 = L_0 L_2 (dt)^2.$$

This result indicates the degree of generality which attaches to the geodesic problem on an m-dimensional surface.

4. Normal form (two degrees of freedom). The transformations deduced above admit of particularly elegant application of the case of two degrees of freedom.* In this case the differential element

$$L_2 \, dt^2 = \frac{1}{2}(a_{11}\, dq_1^2 + 2\, a_{12}\, dq_1\, dq_2 + a_{22}\, dq_2^2)$$

may be regarded as the squared element of arc length of a certain two-dimensional surface. By choosing $\overline{q}_1$ and $\overline{q}_2$ to be the coördinates of an isothermal net on the surface, the squared element of arc is given the form

$$\frac{1}{2}\lambda(d\overline{q}_1^2 + d\overline{q}_2^2).$$

Consequently if we choose the function μ as $1/\lambda$, and make the transformation of t above, λ reduces to 1.

For a given Lagrangian system with two degrees of freedom and given energy constant 0, *there exist variables of the above type for which the principal function L has the form*

$$L = \frac{1}{2}(q_1'^2 + q_2'^2) + \alpha q_1' + \beta q_2' + \gamma.$$

The equations and condition then take the normal form

$$q_1'' + \lambda q_2' = \partial\gamma/\partial q_1, \quad q_2'' - \lambda q_1' = \partial\gamma/\partial q_2$$
$$(\lambda = \partial\alpha/\partial q_2 - \partial\beta/\partial q_1),$$
$$\frac{1}{2}(q_1'^2 + q_2'^2) = \gamma.$$

* See my paper *Dynamical Systems with Two Degrees of Freedom*, Trans. Amer. Math. Soc., vol. 18 (1917), sections 2-5.

Now if we regard q_1, q_2 as the rectangular coördinates of a particle of unit mass in the plane, it is seen that the above equations express the fact that the particle moves subject to a field of force derived from a potential energy $-\gamma$ and a force of magnitude λv perpendicular to the direction of motion, where v denotes velocity.

Any such Lagrangian system with two degrees of freedom can be regarded as that of a mass particle in the q_1, q_2-plane, subject to a conservative field of force derived from a potential energy $-\gamma$, and a non-energic force λv (v, velocity) acting in a direction perpendicular to the direction of motion.

5. Ignorable coördinates. The search for integrals is a task of fundamental importance in connection with differential systems. The question as to whether integrals of a particular type exist or not can usually be answered by formal methods. Their determination has been considered in many cases. In order to refer somewhat to this phase of dynamics, we consider briefly integrals of Lagrangian systems which are either linear or quadratic in the velocities. The variables $q_1, \cdots, q_m$ are confined to the small neighborhood of a point $q_1^0, \cdots, q_m^0$ while $q_1', \cdots, q_m'$ are arbitrary for the integrals treated.

We shall assume that L is quadratic in the velocities with the homogeneous quadratic component L_2 a positive definite form.

There is one very simple case in which a particular integral of the Lagrangian equations linear in the velocities can be found immediately, namely the case in which one of the coördinates, as q_1, does not appear explicitly in the principal function L. In this case, the corresponding differential equation becomes

$$\frac{d}{dt}\left(\frac{\partial L}{\partial q_1'}\right) = 0,$$

so that

$$\partial L / \partial q_1' = c$$

is an integral linear in the velocities. The coördinate q_1 is then said to be an 'ignorable coördinate'.

It may be proved by the method of variation that the $m-1$ equations remaining, which give a system of $m-1$ equations of the second order in $q_2, \cdots, q_m$ after the above integral has been used to eliminate q_1', can be expressed in Lagrangian form. Let us denote by $\overline{L}$ the function of $q_2, \cdots, q_m, q_2', \cdots, q_m'$ obtained from L by this elimination. If $q_1^0, \cdots, q_m^0$ satisfy the given Lagrangian equations, we find for an arbitrary variation of $q_2, \cdots, q_m$

$$\delta \int_{t_0}^{t_1} \overline{L}\, dt = \sum_{j=1}^{m} \frac{\partial L}{\partial q_j'} \delta q_j \Big|_{t_0}^{t_1}$$

after an integration by parts; here q_1' is determined by the integral relation, although q_1 is not determined up to an additive constant. If the $\delta q_2, \cdots, \delta q_m$ vanish near the end points, this reduces to

$$\delta \int_{t_0}^{t_1} \overline{L}\, dt = c\, \delta q_1 \Big|_{t_0}^{t_1} \quad \text{or} \quad \delta \int_{t_0}^{t_1} (\overline{L} - c q_1')\, dt = 0.$$

If q_1 is an ignorable coördinate, the Lagrangian equations can be replaced by a set of Lagrangian equations in $q_2, \cdots, q_m$ only, with modified principal function

$$L - \frac{\partial L}{\partial q_1'} q_1',$$

in which the known integral is used to eliminate q_1'.

We sketch the above reduction of the number of degrees of freedom by use of such an integral because it is typical of the kind of reduction aimed at in many dynamical problems, namely a reduction maintaining the general form of the equations.

6. The method of multipliers. Let us ask next the following question: Under what conditions is it possible to find m 'multipliers' M_i, depending upon the coördinates and the velocities, such that when the Lagrangian equations are multiplied by $M_1, \cdots, M_m$ respectively and added, the left-hand member of the resulting equation is the exact derivative of a function V linear in the velocities? If a set of such multipliers exist, we have

$$\sum_{j=1}^{m} M_j \left[\frac{d}{dt}\left(\frac{\partial L}{\partial q_j'}\right) - \frac{\partial L}{\partial q_j} \right] = \frac{dV}{dt}.$$

Evidently this will lead to a generalization of the notion of ignorable coördinates, in which special case we have $M_i = 1$ for some i while $M_j = 0$ for $i \neq j$.

On comparing coefficients of q_i'' we derive first

$$\sum_{j=1}^{m} M_j \frac{\partial^2 L}{\partial q_i' \, \partial q_j'} = \frac{\partial V}{\partial q_i'} \quad (i = 1, \cdots, m).$$

Here, because of the assumption on L, the coefficients of M_j are functions of the coördinates only. The right-hand member is also a function of the coördinates only, since V is linear in the q_i'. Hence the functions M_i must involve only the coördinates, and partial integration with respect to q_i' yields

$$V = \sum_{j=1}^{m} M_j \frac{\partial L}{\partial q_j'} + S(q_1, \cdots, q_m).$$

For a given V only one such a set of functions M_i, S exist, since the coefficients $\partial L / \partial q_j'$ of M_j are linearly independent expressions in the velocities, q_i'. Furthermore, this type of relation will persist if the variables are changed, since an integral linear in the velocities remains linear under any change of variable. Making then, a change from q_i to $\overline{q}_i$, we find

$$V = \sum_{j,k=1}^{m} M_j \frac{\partial L}{\partial \overline{q}_k'} \frac{\partial \overline{q}_k}{\partial q_j} + S.$$

Thus the new coefficients are given by

$$\overline{M}_i = \sum_{j=1}^{m} M_j \frac{\partial \overline{q}_i}{\partial q_j}.$$

From the known theory of linear partial differential equations of the first order, we can determine m functionally independent functions $\bar{q}_i$ such that we have the relations:

$$\bar{M}_1 = 1, \qquad \bar{M}_2 = \cdots = \bar{M}_m = 0.$$

On making this change, we obtain

$$V = \partial L / \partial \bar{q}_1' + S.$$

Differentiating with respect to t, and using the first Lagrangian equation, we find the identity

$$\frac{\partial L}{\partial q_1} + \sum_{j=1}^{m} \frac{\partial S}{\partial q_j} q_j' \equiv 0.$$

Hence $\partial L / \partial q_1$ is linear in the velocities. Consequently the quadratic terms in L must have the form

$$L_2 = \sum_{j,k=1}^{m} a_{jk}(q_2, \cdots, q_m)\, q_j' q_k'.$$

Now let us write

$$L_1 = \sum_{j=1}^{m} b_j(q_1, \cdots, q_m)\, q_j', \qquad L_0 = e(q_1, \cdots, q_m).$$

Then the above identity simplifies to

$$\sum_{j=1}^{m} \frac{\partial b_j}{\partial q_1} q_j' + \frac{\partial e}{\partial q_1} + \sum_{j=1}^{m} \frac{\partial S}{\partial q_j} q_j' = 0.$$

We infer at once that e is independent of q_1, and that if we write $S^* = \int S\, dq_1$, then L_1 is given by

$$L_1 = -\sum_{j=1}^{m} \frac{\partial S^*}{\partial q_j} q_j' + \sum_{j=2}^{m} b_j^*\,(q_2, \cdots, q_m)\, q_j',$$

i. e., by an exact differential augmented by a linear expression in $q_2', \cdots, q_m'$ with coefficients depending only upon $q_2, \cdots, q_m$.

Since the L function may be modified by an exact derivative without affecting the variation and the Lagrangian equations, we may omit the first term in L_1. Hence L may be written so as not to involve the coördinate q_1 directly.

The most general case, in which multipliers $M_i(q_1, \cdots, q_m)$ *of the various Lagrangian equations exist, by the aid of which the left-hand members of these equations may be combined to form an exact derivative of a function V linear in the velocities* $q_1', \cdots, q_m'$, *reduces by change of variable to the case of an ignorable coördinate* q_1 *in which all of the multipliers but one are zero and that one is unity.*

The existence of such linear integrals can be determined by purely geometric methods. We observe that in the derivation of the result above, only transformations involving $q_1, \cdots, q_m$ were made so that t was unchanged. Hence the quadratic differential form $ds^2 = L_2\, dt^2$ is an invariant, which in the final variables has coefficients only involving $q_2, \cdots, q_m$. But of course this analytic property merely means that the surface with differential element belonging to this form admits of one-parameter continuous group of transformations into itself,

$$\overline{q}_1 = q_1 + c,\ \overline{q}_2 = q_2,\ \cdots,\ \overline{q}_m = q_m.$$

A necessary condition for the existence of such a generalized ignorable coördinate is that the surface $ds^2 = L_2\, dt^2$ *admits of a one-parameter continuous group of transformations into itself.*

We shall not attempt to develop such necessary conditions further.

7. The general integral linear in the velocities. So far as our reasoning above is concerned, we cannot as yet infer that all integrals linear in the velocities can be obtained by the method of generalized ignorable coördinates. However, this may be demonstrated to be the case as follows.

Since L_2 is by assumption a positive definite form, we may write the integral in the form used in the preceding section,

$$V = \sum_{j=1}^{m} M_j \frac{\partial L}{\partial q_j'} + S$$

where M_i and S are functions of the coördinates only. Employing exactly the method of that section, it appears that by a suitable change of variables we can take $M_1 = 1$, $M_2 = \cdots = M_m = 0$, and then by differentiation as to t it appears just as there, that L is essentially independent of q_1, so that q_1 is ignorable.

The method of multipliers specified yields all integrals of the Lagrangian equations which are linear in the velocities.

8. Conditional integrals linear in the velocities. In the preceding section we have considered integrals linear in the velocities which hold for *all* values of the energy constant. A more difficult problem is that of obtaining the conditional integral, holding for a *specified* particular value of the energy constant c, say for $c = 0$. We proceed to treat this problem for the case of two degrees of freedom. Here, by the use of the normalizing variables obtained earlier, we may write the equations of motion and the energy integral in the form:

$$x'' + \lambda y' = \gamma_x, \quad y'' - \lambda x' = \gamma_y; \quad x'^2 + y'^2 = \gamma,$$

where γ_x, for instance, denotes $\partial \gamma / \partial x$.

Moreover, since any change of variables leaves the linear nature of the integral unaltered, the integral may be written

$$V = lx' + my' + n = k,$$

where it is understood that this relation is required to hold only when the energy constant vanishes.

If the linear integral be differentiated as to the time, the equation which results must be an identity in virtue of the differential equations of motion written above and the energy relation. The differential equations may be employed to eliminate x'', y''. When this has been done, an equation quadratic in x', y' is obtained, which must be an identity in virtue of the integral relation alone. The quadratic terms are

$$l_x x'^2 + (l_y + m_x)\, x'\, y' + n_y y'^2.$$

In order that this sum shall combine with those of lower degrees in x', y' by use of the integral relation, it must be of the form $\varrho(x'^2 + y'^2)$. This implies

$$l_x = m_y, \qquad l_y = -m_x$$

i. e., that

$$l = u_y, \qquad m = u_x$$

where u is a harmonic function.

The integral can now be written:

$$u_y x' + u_x y' + n = k.$$

According to the principles outlined above in section 4, a further arbitrary conformal transformation of the x, y-plane, joined with the appropriate change in t, will leave the differential equation and integral relation in the normal form. In order to simplify further the linear integral, we shall choose the transformation to $\bar{x}$, $\bar{y}$ defined by

$$\bar{x} + i\bar{y} = \int \frac{dx + i\,dy}{u_y + i\,u_x} \qquad (i = \sqrt{-1}).$$

This is evidently conformal in type. The inverse transformation,

$$x + iy = f(\bar{x} + i\bar{y}),$$

is also conformal, and we have

$$|f'(\bar{x} + i\bar{y})|^2 = \left| \frac{dx + i\,dy}{d\bar{x} + i\,d\bar{y}} \right|^2 = u_y^2 + u_x^2.$$

Now let the transformed value of t be defined by

$$dt = (u_y^2 + u_x^2)\, d\bar{t}$$

From this last equation we find at once

$$\bar{x}' + i\bar{y}' = (u_y - iu_x)(x' + iy')$$

where $\bar{x}' = d\bar{x}/d\bar{t}$, $\bar{y}' = d\bar{y}/d\bar{t}$. Thus we have in particular

$$\bar{x}' = u_y x' + u_x y'.$$

Consequently when such a further transformation has been made, the above integral is simplified to

$$x' + n = k.$$

Now let this integral be differentiated as to t and let x'' be eliminated by means of the first Lagrangian equation. There results

$$n_x x' + (n_y - \lambda) y' + \gamma_x = 0,$$

which must vanish identically in virtue of the integral relation. Therefore, we conclude that the left-hand member vanishes identically in x', y'. But this will happen only if λ and γ are functions of y only. In this case the equation can be made to vanish identically by a proper choice of n, namely $\int \lambda\, dy$.

If such a dynamical system with two degrees of freedom with energy constant 0 *admits of a conditional integral linear in the velocities, then by means of a suitable transformation of the coördinates and the time, the equations can be taken in normal form with*

$$L = \frac{1}{2}(x'^2 + y'^2) + n(y)x' + \gamma(y),$$

so that the system contains the ignorable coördinate x. In this integrable case the curves of motion are given by

$$x = \int \frac{(c_1 - n)\, dy}{\sqrt{2\gamma - (c_1 - n)^2}} + c_2,$$

$$t = \int \frac{dy}{\sqrt{2\gamma - (c_1 - n)^2}} + c_3.$$

9. Integrals quadratic in the velocities. The energy integral is a known integral which is quadratic in the velocities. Furthermore it is well known that dynamical systems of the so-called Liouville type with L of the form

$$L = \frac{1}{2} U \sum_{j=1}^{m} v_j(q_j) q_j'^2 - (W/U),$$

$$U = \sum_{j=1}^{m} u_j(q_j), \qquad W = \sum_{j=1}^{m} w_j(q_j)$$

admit of m integrals quadratic in the velocities, in particular

$$\frac{1}{2} U^2 v_i q_i'^2 - c u_i + w_i = c_i \qquad (i = 1, \cdots, m),$$

and can be completely integrated.

We propose here only to discuss a special converse problem: to determine the conditions under which a Lagrangian system with two degrees of freedom and of reversible type, with energy constant 0, admits of a conditional integral

$$\frac{1}{2}(a x'^2 + 2 b x' y' + c y'^2) + d x' + e y' + f = k$$

where $a, \cdots, f$ are functions of x and y, and where a, b, c are not all identically zero.

If such an integral exists, any transformation of x, y, t of the type discussed in section 3 leaves the form of the integral unaltered. Hence we may transform the equations to the normal form for which

$$L = \frac{1}{2}(x'^2 + y'^2) + \gamma.$$

Differentiating the above assumed integral relation, and making use of the Lagrangian equations to eliminate x'', y'', we obtain a polynomial of the third degree in x', y' at most, which must vanish identically in virtue of the above integral relation. Now the third degree terms are

$$\frac{1}{2} a_x x'^3 + \left(b_x + \frac{1}{2} a_y\right) x'^2 y' + \left(b_y + \frac{1}{2} c_x\right) x' y'^2 + \frac{1}{2} c_y y'^3,$$

and these must combine with those of lower degree by virtue of the integral relation. This can only happen if this polynomial is divisible by $x'^2 + y'^2$, i. e., if

$$a_x - c_x = 2\,b_y, \qquad a_y - c_y = -2\,b_x.$$

These are the Cauchy-Riemann differential equations for the conjugate harmonic functions $a - c$, $2b$, and we may write

$$a - c = 2\,u_y, \qquad b = u_x,$$

where u is a harmonic function.

Our conclusion is that the hypothetical integral has quadratic terms

$$\frac{1}{2} u_y\, x'^2 + u_x\, x'\, y' - \frac{1}{2} u_y\, y'^2 + \varrho\,(x'^2 + y'^2).$$

Taking account of the energy relation we may replace the last term by $2\,\varrho\gamma$. The remaining quadratic terms may be written

$$\frac{1}{2} \Re\, [(u_y - i\,u_x)\ (x' + iy')^2]$$

where $\Re$ stands for 'the real part of'.

Now write

$$f'^2 = 1/(u_y + i u_x)$$

so that f is an analytic function of $x + iy$. Make the change of variables

$$\bar{x} + i\bar{y} = f(x + iy), \qquad d\bar{t} = |f'|^2\, dt,$$

which leaves the normal form of the equations unaltered. We find that the above quadratic terms, which may be written

$$\frac{1}{2} \Re \left[\frac{f'^2\,(dx + i\,dy)^2}{|f'|^4\, dt^2} \right]$$

become

$$\frac{1}{2} (\bar{x}'^2 - \bar{y}'^2)$$

in the new variables. Hence, dropping the bars, the integral relation takes the simplified form

$$\frac{1}{2}(x'^2 - y'^2) + dx' + ey' + f = k.$$

Again if this be differentiated with respect to t as before, there is obtained

$$\begin{aligned} d_x x'^2 + (d_y + e_x)\, x' y' + e_y y'^2 \\ + (f_x + \gamma_x)\, x' + (f_y - \gamma_y)\, y' + d\gamma_x + e\gamma_y = 0. \end{aligned}$$

The linear terms must vanish so that we find

$$\gamma = \varphi(x) + \psi(y), \qquad f = -\varphi(x) + \psi(y).$$

But for this value of γ the differential equations are of immediately integrable type:

If a reversible Lagrangian system with two degrees of freedom and with the energy constant 0 *admits of a conditional integral quadratic in the velocities and distinct from the energy integral, then, by a transformation of variables, the equations and integral take the form*

$$x'' = \varphi'(x), \qquad y'' = \psi'(y), \qquad \frac{1}{2}(x'^2 + y'^2) = \varphi(x) + \psi(y).$$

A special quadratic integral is then

$$\frac{1}{2}(x'^2 - y'^2) = \varphi(x) - \psi(y) + k$$

and the equations are integrable with

$$t = \frac{1}{\sqrt{2}} \int \frac{dx}{\sqrt{\varphi + k}} = \frac{1}{\sqrt{2}} \int \frac{dy}{\sqrt{\psi - k}}.$$

The Liouville type of equations is essentially an equivalent case.

10. The Hamiltonian equations. Next we proceed to formulate another important type of variational principle, which leads to the so-called Hamiltonian or canonical form of the equations of dynamics.

Let us write

$$\delta\int_{t_0}^{t_1}\left[\sum_{j=1}^{m} p_j q_j' - \sum_{j=1}^{m} p_j r_j + L(q_1, \cdots, q_m, r_1, \cdots, r_m)\right] dt = 0,$$

in which the r_i are the functions of $p_1, \cdots, p_m, q_1, \cdots, q_m$ properly defined by the m equations

$$p_i = \partial L/\partial r_i \qquad (i = 1, \cdots, m),$$

and where $p_1, \cdots, p_m, q_1, \cdots, q_m$ are to be varied independently. The first m equations, obtained from the variation of $p_1, \cdots, p_m$, are of course

$$\frac{d}{dt}\left(\frac{\partial F}{\partial p_i'}\right) - \frac{\partial F}{\partial p_i} = -q_i' + r_i + \sum_{j=1}^{m}\left(p_j \frac{\partial r_j}{\partial p_i} - \frac{\partial L}{\partial r_j}\,\frac{\partial r_j}{\partial p_i}\right)$$
$$= -q_i' + r_i = 0,$$

where F stands for the integrand. The second set of m equations can be likewise obtained and may be written

$$p_i' + \partial H/\partial q_i = 0,$$

if we introduce the abbreviation H for

$$\sum_{j=1}^{m} p_j r_j - L.$$

It is important to observe that the $2m$ differential equations so obtained are each only of the first order, with the general solution containing only $2m$ arbitrary constants.

The first set of equations show that the functions p_i^0, q_i^0 which make the integral stationary are such that $r_i^0 = q_i^{0\prime}$. Now let r_i be fixed as q_i', so that the integral reduces to the Lagrangian integral

$$\int_{t_0}^{t_1} L(q_1, \cdots, q_m, q_1', \cdots q_m')\, dt.$$

The variation of $q_1, \cdots, q_m$ is still arbitrary, but the variation of $p_1, \cdots, p_m$ is determined. Furthermore if the variations of $q_1, \cdots, q_m$ vanish near to t_0 and t_1, so will the variations of $p_1, \cdots, p_m$. Hence we have

$$\delta \int_{t_0}^{t_1} L\,dt = 0,$$

along $q_i = q_i^0(t)$, and we conclude that q_i^0 satisfy the associated Lagrangian equations, with $r_i^0 = q_i^{0\prime}$, thus determining the corresponding p_i^0.

Thus each solution of the proposed variational problem leads to a solution of the associated Lagrangian equations. The converse is also true, since the choice of p_i, q_i at any time t is arbitrary and leads to an arbitrary set of values of q_i, q_i'.

If the principal function for a Lagrangian system is $L(q_1, \cdots, q_m, q_1', \cdots, q_m')$ *and we form the function of* $p_1, \cdots, p_m, q_1, \cdots, q_m$ *defined by*

$$H = -L + \sum_{j=1}^{m} p_j q_j', \tag{3}$$

where the variables q_i' *are to be eliminated by means of the equations*

$$p_i = \partial L / \partial q_i' \qquad (i = 1, \cdots, m), \tag{4}$$

the original equations $\delta \int L\,dt = 0$ *may be replaced by the equivalent system in* p_i, q_i

$$\delta \int_{t_0}^{t_1} \left[\sum_{j=1}^{m} p_j q_j' - H \right] dt = 0, \tag{5}$$

or, more explicitly,

$$dp_i/dt = -\partial H/\partial q_i, \quad dq_i/dt = \partial H/\partial p_i \quad (i = 1, \cdots, m). \tag{6}$$

The equations (6) are the 'Hamiltonian' equations, and the variables p_i are called the 'generalized momenta'. A pair of variables p_i, q_i are called 'conjugate'. Furthermore it is to be noted that the Hamiltonian 'principal function' H is the total energy expressed in terms of the generalized coördinates and momenta. The energy integral $H =$ const. follows at once from the canonical equations.

It may be observed here that the above variational principle leads to the same canonical equations even if L and H involve the time t.

Conversely, any Hamiltonian system (5), (6), *H being arbitrary, can be reduced to a Lagrangian system.*

To prove this statement we need only define L by the equation

$$L(q_1, \cdots, q_m, r_1, \cdots, r_m) = -H + \sum_{j=1}^{m} p_j r_j$$

where $p_1, \cdots, p_m$ are functions of q_i and r_i given by the implicit relations

$$r_i = \partial H / \partial p_i \qquad (i = 1, \cdots, m).$$

It is obvious that the Lagrangian system with this principal function L is associated with the prescribed function H in the way desired.

If H contains t, so will L of course, and the same method is applicable.

11. Transformation of the Hamiltonian equations. The variational principle (5) is remarkable in that it only involves the second half of the derivatives $p_1', \cdots, p_m'$, $q_1', \cdots, q_m'$ under the integral sign, and those linearly with coefficients precisely the conjugate variables. A general point transformation from $p_1, \cdots, q_m$ to $\bar{p}_1, \cdots, \bar{q}_m$ will yield a form linear in $p_1', \cdots, q_m'$ but not of this special type. We shall desire in the next section to consider the corresponding Pfaffian type of equation so obtained, which has certain advantages over the Hamiltonian type.

A general 'contact transformation' preserving the canonical form is the following

$$(7) \qquad p_i = \partial K / \partial q_i, \; \bar{p}_i = -\partial K / \partial \bar{q}_i \qquad (i = 1, \cdots, m),$$

where K is an arbitrary function of $q_1, \cdots, q_m, \bar{q}_1, \cdots, \bar{q}_m, t$ except it must be such as to define a proper transformation from $p_1, \cdots, q_m$ to $\bar{p}_1, \cdots, \bar{q}_m$ by means of the above equations. We shall not undertake to explain the apparent artificiality in these equations, but proceed to prove that such transformations do indeed leave invariant the canonical form. By use of the first m of these equations we modify the variational problem to the form

$$\delta \int_{t_0}^{t_1} \left[\sum_{j=1}^{m} \frac{\partial K}{\partial q_j} q_j' - H \right] dt = 0$$

where the independent variables are now taken as $\overline{p}_1, \cdots, \overline{q}_m$. But for these same variables, we have

$$\delta \int_{t_0}^{t_1} \left[\sum_{j=1}^{m} \left(\frac{\partial K}{\partial q_j} q_j' + \frac{\partial K}{\partial \overline{q}_j} \overline{q}_j' + \frac{\partial K}{\partial t} \right) \right] dt = 0,$$

since the expression under the integral sign is an exact derivative. By subtraction and use of the second set of m equations of transformation we deduce

$$\delta \int_{t_0}^{t_1} \left[\sum_{j=1}^{m} \overline{p}_j \overline{q}_j' - \overline{H} \right] dt = 0 \quad (\overline{H} = H + \partial K / \partial t).$$

The transformation (7) *preserves the Hamiltonian form with* $\overline{H} = H + \partial K / \partial t$, *in case the arbitrary function* K *yields a proper transformation.*

Similarly we may write

$$(8) \qquad p_i = \partial K / \partial q_i, \qquad \overline{q}_i = \partial K / \partial \overline{p}_i \quad (i = 1, \cdots, m),$$

and find a corresponding result.

The transformation (8) *also preserves the Hamiltonian form with* $\overline{H} = H + \partial K / \partial t$.

It deserves to be remarked that transformations of type (8) form a group. In fact such a transformation is characterized by the fact that

$$\sum_{j=1}^{m} (p_j \, dq_j + \overline{q}_j \, d\overline{p}_j)$$

is an exact differential, dK. For a second such transformation from $\overline{p}_1, \cdots, \overline{q}_m$ to $\overline{\overline{p}}_1, \cdots, \overline{\overline{q}}_m$, there is a second characteristic $d\overline{K}$. By addition we infer

$$\sum_{j=1}^{m} (p_j \, dq_j + \bar{p}_j \, d\bar{q}_j) = d(K + \bar{K} - \sum_{j=1}^{m} \bar{p}_j \, d\bar{q}_j),$$

so that the compound transformation is of the same type. Similarly the inverse of a transformation (7), or the resultant of an odd number of transformations is of the same type, while the resultant of an even number of transformations (7) is of type (8).*

12. The Pfaffian equations. It is clear that Hamiltonian equations can be regarded as a special type arising from the more general Pfaffian variational principle,

$$(9) \qquad \delta \int_{t_0}^{t_1} \left[\sum_{j=1}^{n} P_j p_j' + Q \right] dt = 0,$$

in which the integral is linear in all of the first derivatives with arbitrary functions $P_1, \cdots, P_n$, Q of $p_1, \cdots, p_n$ as coefficients, and n is even.

If we develop these equations explicitly they become

$$(10) \qquad \sum_{j=1}^{n} \left(\frac{\partial P_i}{\partial p_j} - \frac{\partial P_j}{\partial p_i} \right) \frac{dp_j}{dt} - \frac{\partial Q}{\partial p_i} = 0 \quad (i = 1, \cdots, n).$$

Furthermore these equations are evidently those of a degenerate Lagrangian problem with $L_2 = 0$, $L_1 = \sum P_j p_j'$, $L_0 = Q$, so that there is the particular integral $Q =$ const. This reduces to the energy integral in the Hamiltonian case.

These equations admit of an arbitrary point transformation of all of the variables without losing their form. It is only necessary to determine the modified linear differential form under the integral sign by direct substitution. Thus the Pfaffian equations admit of perfect flexibility of transformation, and in this respect are easier to deal with than either the Lagrangian or Hamiltonian equations.

13. On the significance of variational principles. Since the variational principles have taken an important

* For the applications of the theory of contact transformations and for consideration of the associated Hamiltonian partial differential equation, the reader is referred to Whittaker, *Analytical Dynamics*, chaps. 10, 11, 12.

part in dynamical theory, it is of especial interest to determine their real significance for dynamics. In other words, what especial properties are possessed by the Lagrangian, Hamiltonian or Pfaffian equations arising from the respective variational principles treated above? All of these can be regarded as systems of $n = 2m$ equations of the first order if we introduce the new variables $r_i = q_i'$ in the Lagrangian equations.

Let us first remark that so long as these equations are considered in the vicinity of a point in the corresponding space of n dimensions not an equilibrium point, there are no especial characteristics to be found.

Indeed if we take a dynamical system as defined by any set of n equations

$$dx_i/dt = X_i(x_1, \cdots, x_n) \quad (i = 1, \cdots, n),$$

it will in general remain of the same type under an arbitrary point transformation

$$x_i = \varphi_i(y_1, \cdots, y_n) \quad (i = 1, \cdots, n)$$

under certain conditions. Two systems of this kind will naturally be termed 'equivalent' if it is possible to pass from one to the other by an admissible point transformation of this kind. If we confine attention to the neighborhood of a point $x_1^0, \cdots, x_n^0$ at which not all of the X_i vanish, so that this is not an equilibrium point, the equivalence with other such systems is unrestricted, and the new equations may be taken to be

$$dy_1/dt = 1, \quad dy_i/dt = 0 \quad (i = 2, \cdots, n),$$

for instance. This is readily seen as follows. Conceive of the given differential system as defining a steady fluid motion in $x_1, \cdots, x_n$ space so that the curves of motion are defined by the solution $x_i = x_i(t)$, $(i = 1, \cdots, n)$. These curves which have a definite direction with direction cosines proportional to $X_1, \cdots, X_n$ may be deformed into the straight lines

$$y_1 = t, \qquad y_2 = c_2, \cdots, y_n = c_n$$

of a $y_1, \cdots, y_n$ space by one-to-one analytic deformation. Consequently the transformed equations have as general solution

$$y_1 = t + c_1, \qquad y_2 = c_2, \cdots, y_n = c_n,$$

whence it follows immediately that these equations have the desired normal form.

Hence in such a domain there is no distinction between equations derived from a variational principle and the most general equation.

In the following chapter we shall see that variational principles play an important role in connection with the formal stability of dynamical systems near equilibrium or periodic motion. Indeed this appears to be their principal significance for dynamics.

One further interesting remark concerning variational principles may be made here. Suppose that we start with n arbitrary equations of the form

$$dx_i/dt = X_i(x_1, \cdots, x_n, t) \qquad (i = 1, \cdots, n). \tag{11}$$

The equations of variation are

$$\frac{dy_i}{dt} = \sum_{j=1}^{n} \frac{\partial X_i}{\partial x_j} y_j \qquad (i = 1, \cdots, n).$$

There can be formally integrated at once if the general solution

$$x_i = f_i(t, c_1, \cdots, c_n) \qquad (i = 1, \cdots, n)$$

is at hand, namely

$$y_i = k_1 \frac{\partial f_i}{\partial c_1} + \cdots + k_n \frac{\partial f_i}{\partial c_n} \qquad (i = 1, \cdots, n)$$

where $k_1, \cdots, k_n$ are arbitrary constants.

Similarly the adjoint system to the equations of variation

$$\frac{dz_i}{dt} = -\sum_{j=1}^{n} \frac{\partial X_j}{\partial x_i} z_j \qquad (i = 1, \cdots, n) \tag{12}$$

can be integrated explicitly by taking

$$\frac{\partial f_1}{\partial c_i} z_1 + \cdots + \frac{\partial f_n}{\partial c_i} z_n = k_i \quad (i = 1, \cdots, n).$$

Hence the given system (11) of equations of the first order can be called 'equivalent' to that of the extended system (11), (12) of twice the order in the $2n$ variables $x_1, \cdots, x_n, z_1, \cdots, z_n$, since the explicit solution of either system involves that of the other. But the extended system (11), (12) is Hamiltonian with conjugate variables x_i, z_i, and with

$$H = -\sum_{j=1}^{m} X_j z_j,$$

as may be directly verified.

These remarks serve to indicate the care necessary in assigning to the variational principles their true significance.

CHAPTER III

FORMAL ASPECTS OF DYNAMICS

1. Introductory remarks. In the preceding chapter it was pointed out that a system of ordinary differential equations

$$(1) \qquad dx_i/dt = X_i(x_1, \cdots, x_n) \quad (i = 1, \cdots, n)$$

was devoid of invariantive characteristics under the general group of one-to-one, analytic transformations

$$(2) \qquad x_i = \varphi_i(\bar{x}_1, \cdots, \bar{x}_n) \qquad (i = 1, \cdots, n),$$

provided that attention be confined to the vicinity of any point x_i^0. This was done under the assumption that the functions X_i were analytic and did not all vanish at the point x_i^0.

The simplest case in which invariantive characteristics can be expected to arise is that of a point of equilibrium, the condition for which is $X_i^0 = 0$, $(i = 1, \cdots, n)$. Another important case, which may be regarded as including this one, is that associated with the neighborhood of a periodic solution of (1) of period τ

$$x_i = f_i(t) \qquad (i = 1, \cdots, n).$$

If we write then

$$x_i = f_i(t) + \bar{x}_i \qquad (i = 1, \cdots, n),$$

the equations take the more general form

$$(3) \qquad dx_i/dt = X_i(x_1, \cdots, x_n, t) \quad (i = 1, \cdots, n),$$

where the functions X_i are analytic in $x_1, \cdots, x_n$ and t, periodic in t with the period τ of the motion, and vanish at the origin in the new $x_1, \cdots, x_n$ space for all values of t. We shall treat the question of the invariantive characteristics for the 'generalized equilibrium problem' defined by a system of this more general type.

In the present chapter we shall in the main restrict attention to those purely formal properties which have no regard to the convergence or divergence of the series employed, and which take for granted an equilibrium point or a known periodic motion. By doing so we shall be able to develop to a considerable extent the formal significance of the Lagrangian, Hamiltonian and Pfaffian types of equations employed in dynamics. In order to do this we propose first of all to develop the characteristics of what may be called the general case of the equilibrium problem, and then to pass on to the more special types referred to, so that a comparison may be effected.

2. The formal group. As a matter of notational convenience, the equilibrium point of (3) will be kept at the origin in all cases. The type of transformation considered will be

$$x_i = \sum_{j=1}^{n} a_{ij}(t)\,\bar{x}_j + \frac{1}{2}\sum_{j,k=1}^{n} a_{ijk}(t)\,\bar{x}_j\,\bar{x}_k + \cdots \tag{4}$$

where the real coefficients $a_{ij}, a_{ijk}, \cdots$ are periodic analytic functions of t with period τ, such that the determinant $|a_{ij}|$ is not zero for any t. Evidently two of these transformations performed successively may be united into a single composite transformation, while the inverse of such a transformation is another of the same type. Furthermore, the form (3) of the differential equations is clearly maintained under this group of transformations.

Imagine now that divergent series appear in (4). The right-hand members $\bar{X}_i$ of the transformed differential equations will then be given as definite formal power series in $\bar{x}_1, \cdots, \bar{x}_n$ with coefficients analytic in t and of period τ, and these series will lack constant terms. Thus along with the formal group we obtain corresponding formal differential equations. Now it is to be particularly stressed that the ordinary laws for composition of transformations, and for deriving the associated differential equations, hold in the case of divergence as well as in that of convergence. This is an obvious consequence of the fact that when the formal series involved are broken off at terms of high degree, actual transformations and actual differential equations are obtained for which these formal laws are valid. By including terms of higher and higher degree the lower degree terms in the differential equations are not affected. Proceeding thus to the formal limiting case, we infer that the usual purely formal relations will continue to hold in the case of divergence.

In many cases it is convenient to introduce a slight extension of the above formal group so as to take care of certain pairs of variables x_i and x_j in a special way. In fact it is convenient to introduce conjugate variables

$$\xi = x_i + \sqrt{-1}\, x_j, \qquad \eta = x_i - \sqrt{-1}\, x_j$$

so that if x_i and x_j are real, ξ and η are conjugate imaginaries, and conversely. At the same time the transformation from $x_1, \cdots, x_n$ to $\bar{x}_1, \cdots, \bar{x}_n$ may be expressed in terms of the conjugate pairs such as ξ, η and the corresponding transformed variables $\bar{\xi}, \bar{\eta}$. The series involved are then characterized by the property that if the conjugate variables ξ and η are given conjugate imaginary values, while those not so paired are real, then the same will hold for the new variables. If we go back to the underlying transtormation belonging to the formal group, the new variables $\bar{x}_1, \cdots, \bar{x}_n$ will be real if $x_1, \cdots, x_n$ are.

It is not difficult to determine the characteristics of the formal series which appear in the transformation of such conjugate pairs of variables. If the transformation within the original group be written

$$x_i = f_i(\bar{x}_1, \bar{y}_1, \cdots, \bar{x}_s, \bar{y}_s, \bar{x}_{2s+1}, \cdots, \bar{x}_n),$$
$$y_i = g_i(\bar{x}_1, \bar{y}_1, \cdots, \bar{x}_s, \bar{y}_s, \bar{x}_{2s+1}, \cdots, \bar{x}_n),$$
$$(i = 1, \cdots, s),$$
$$x_i = h_i(\bar{x}_1, \bar{y}_1, \cdots, \bar{x}_s, \bar{y}_s, \bar{x}_{2s+1}, \cdots, \bar{x}_n),$$
$$(i = 2s+1, \cdots, n),$$

where $x_1, y_1, \cdots, x_s, y_s$ are the s pairs of associated variables, then the conjugate variables are

$$\xi_i = x_i + \sqrt{-1}\, y_i, \quad \eta_i = x_i - \sqrt{-1}\, y_i \quad (i = 1, \cdots, s)$$

with like definitions of $\bar{\xi}_i, \bar{\eta}_i$ in terms of $\bar{x}_i, \bar{y}_i$. Hence we have explicitly

$$\xi_i = f_i((\bar{\xi}_1 + \bar{\eta}_1)/2, (\bar{\xi}_1 - \bar{\eta}_1)/2\sqrt{-1}, \cdots, \bar{x}_{2s+1}, \cdots, \bar{x}_n)$$
$$+ g_i((\bar{\xi}_1 + \bar{\eta}_1)/2, (\bar{\xi}_1 - \bar{\eta}_1)/2\sqrt{-1}, \cdots, \bar{x}_{2s+1}, \cdots, \bar{x}_n)\sqrt{-1}$$

for $i = 1, \cdots, s$ with like formulas for $\eta_i, (i = 1, \cdots, s)$ and for $x_i, (i = 2s + 1, \cdots, n)$, when we employ the modified variables. Clearly the series on the right have the properties of the transformations of the formal group except that the periodic coefficients are in general complex.

If we examine the form of these series we perceive at once that they posses the following additional characteristic property. If the pairs $\bar{\xi}_i, \bar{\eta}_i$ be interchanged in the series on the right, and if at the same time the periodic coefficients be replaced by their conjugates, then the series for $\xi_i, \eta_i, (i = 1, \cdots, s)$ are interchanged, while those for $x_i, (i = 2s + 1, \cdots, n)$ are unaltered.

It is readily proved that this necessary formal property is also sufficient. In fact suppose that x_i, y_i are real quantities and define $\bar{\xi}_i, \bar{\eta}_i$ as before so that $\bar{\xi}_i, \bar{\eta}_i$ are conjugate complex quantities. Write

$$\xi_i = \varphi_i(\bar{\xi}_1, \cdots, \bar{\eta}_s, \bar{x}_{2s+1}, \cdots, \bar{x}_n, t) \quad (i = 1, \cdots, s),$$

$$\eta_i = \psi_i(\bar{\xi}_1, \cdots, \bar{\eta}_s, \bar{x}_{2s+1}, \cdots, \bar{x}_n, t) \quad (i = 1, \cdots, s),$$

$$x_i = \chi_i(\bar{\xi}_1, \cdots, \bar{\eta}_s, \bar{x}_{2s+1}, \cdots, \bar{x}_n, t) \quad (i = 2s + 1, \cdots, n),$$

where the series $\varphi_i, \psi_i, \chi_i$ are assumed to have the stated formal property and, at the outset, to be convergent. If then we use the $*$ as superscript to denote the 'conjugate of', we find, for instance,

$$\begin{aligned}\xi_i^* &= \varphi_i^*(\bar{\xi}_1^*, \cdots, \bar{\eta}_s^*, x_{2s+1}, \cdots, x_n, t)\\ &= \varphi_i^*(\bar{\eta}_1, \cdots, \bar{\xi}_s, x_{2s+1}, \cdots, x_n, t)\\ &= \psi_i(\bar{\xi}_1, \cdots, \bar{\eta}_s, x_{2s+1}, \cdots, x_n, t) = \eta_i;\end{aligned}$$

here φ_i^* designates the formal series obtained from φ_i by replacing the periodic coefficients by their conjugates. Hence ξ_i and η_i are conjugates also. The same kind of argument shows that x_i must be real for $i = 2s+1, \cdots, n$.

In case the series of the underlying formal gronp are divergent, the series belonging to conjugate variables may also be divergent. But of course the above formal property is still maintained as may be proved by breaking of the formal series at terms of high degree and then applying the above argument.

As a very simple instance of the change from real to conjugate variables let us suppose that the transformation in real form involves one pair of variables $\bar{x}$, $\bar{y}$ as follows:

$$\begin{aligned}x &= \bar{x}\cos(\theta+t-c\bar{r}^2)-\bar{y}\sin(\theta+t-c\bar{r}^2) \qquad (\bar{r}^2 = \bar{x}^2+\bar{y}^2),\\ y &= \bar{x}\sin(\theta+t-c\bar{r}^2)+\bar{y}\cos(\theta+t-c\bar{r}^2).\end{aligned}$$

Evidently this is a transformation of the admitted type with $\tau = 2\pi$. Introducing the corresponding pair of conjugate variables ξ, η we find readily

$$\xi = \bar{\xi}e^{\sqrt{-1}(\theta+t-c\bar{\xi}\bar{\eta})}, \qquad \eta = \bar{\eta}e^{-\sqrt{-1}(\theta+t-c\bar{\xi}\bar{\eta})},$$

and the characteristic property of the series on the right is at once verified.

3. Formal solutions. Suppose that we substitute in the system (3) of differential equations under consideration

$$(5) \qquad x_i = F_i(t, c_1, \cdots, c_n) \qquad (i = 1, \cdots, n),$$

where the F_i are formal power series in the n arbitrary constants $c_1, \cdots, c_n$ without constant terms but such that the determinant $|\partial F_i/\partial c_j|$ is not zero for any t at the origin, and where the coefficients in these series are real and analytic in t; the series may converge or diverge. It may happen that the n equations (3) so obtained will be satisfied identically by these series x_i in the formal sense. In this case we shall say that the series (5) give the 'general formal solution' of the given differential system.

In a special case the coefficients in F_i may happen to be periodic in t of period τ. There would then be defined a corresponding transformation

$$x_i = F_i(t, y_1, \cdots, y_n) \qquad (i = 1, \cdots, n)$$

in the formal group. If the associated formal differential system is

$$dy_i/dt = Y_i(y_1, \cdots, y_n, t) \qquad (i = 1, \cdots, n),$$

we have then the formal identities

$$\frac{\partial F_i}{\partial t} + \sum_{j=1}^{n} \frac{\partial F_i}{\partial y_j} Y_j \equiv X_i(F_1, \cdots, F_n, t) \qquad (i = 1, \cdots, n),$$

in which the arguments of F_i are $y_1, \cdots, y_n, t$ of course. But the precise meaning of the hypothesis that the F_i yield a formal solution is that

$$\partial F_i/\partial t \equiv X_i(F_1, \cdots, F_n, t) \qquad (i = 1, \cdots, n),$$

where now $c_1, \cdots, c_n$ replace $y_1, \cdots, y_n$ as the variables in the series F_i. And if we replace $c_1, \cdots, c_n$ by $y_1, \cdots, y_n$ respectively, as we may, and compare with the equations which precede, it appears that

$$\sum_{j=1}^{n} (\partial F_i/\partial y_j) Y_j \equiv 0,$$

whence of course $Y_i \equiv 0$, $i = 1, \cdots, n$. Consequently this special case is the case in which the given differential system can be formally transformed into the normal form

$$dx_i/dt = 0 \qquad (i = 1, \cdots, n)$$

by a transformation within the specified formal group. As we shall see, this is not in general possible of accomplishment.

If the variables in (3) *are transformed within the formal group, any general formal solution* (5) *is transformed into a general formal solution of the transformed equations.*

An easy way of proving this is to extend the formal group for the moment by allowing analytic coefficients not periodic of period τ. There will then be a transformation, obtained from (5) by replacing $c_1, \cdots, c_n$ by $y_1, \cdots, y_n$ respectively, which is in the extended group, and which takes the given equations into the system for which $Y_i = 0$ $(i = 1, \cdots, n)$ as before. But the transformed system in $\bar{x}_1, \cdots, \bar{x}_n$ can then be taken into this special system directly by means of the composite transformation taking $\bar{x}_i$ to x_i, and then x_i to y_i. But this means precisely that the general solution of the transformed system may be obtained in the specified manner. It is assumed in this reasoning that the formal laws remain valid within the extended formal group.

The most general formal solution $x_i = G_i(t, d_1, \cdots, d_n)$, $(i = 1, \cdots, n)$ *of* (3) *can be obtained from any particular formal solution* (5) *by substitution of* n *arbitrary real series in* $d_1, \cdots, d_n$ *for* $c_1, \cdots, c_n$, *say*

$$c_i = \varphi_i(d_1, \cdots, d_n) \qquad (i = 1, \cdots, n),$$

with the sole proviso that the determinant $|\partial \varphi_i / \partial d_j|$ *is not zero for* $d_1 = \cdots = d_n = 0$.

This almost obvious fact may also be established readily by use of the extended group. The two transformations

$$x_i = F_i(t, z_1, \cdots, z_n), \qquad x_i = G_i(t, w_1, \cdots, w_n) \qquad (i = 1, \cdots, n)$$

take the equations (3) into

$$dz_i/dt = 0, \qquad dw_i/dt = 0 \qquad (i = 1, \cdots, n)$$

respectively. Hence we may write

$$z_i = \varphi_i(t, w_1, \cdots, w_n) \quad (i = 1, \cdots, n)$$

as the transformation taking directly the equations in $z_1, \cdots, z_n$ into the equations in $w_1, \cdots, w_n$. We infer at once that

$$\frac{dz_i}{dt} = \frac{\partial \varphi_i}{\partial t} + \sum_{j=1}^{n} \frac{\partial \varphi_i}{\partial w_j} \frac{dw_j}{dt} = \frac{\partial \varphi_i}{dt} = 0,$$

so that the variable t does not appear in $\varphi_1, \cdots, \varphi_n$, i. e.,

$$z_i = \varphi_i(w_1, \cdots, w_n) \qquad (i = 1, \cdots, n).$$

Thus we obtain the formal identities

$$F_i(t, \varphi_1, \cdots, \varphi_n) \equiv G_i(t, d_1, \cdots, d_n) \quad (i = 1, \cdots, n),$$

provided that we replace $w_1, \cdots, w_n$ by $d_1, \cdots, d_n$ respectively. This is the relationship which we set out to prove. Of course $|\partial \varphi_i / \partial d_j|$ is not 0 for $d_1 = \cdots = d_n = 0$ by virtue of the definition of the extended group.

The question of the existence of formal solutions is immediately disposed of. In fact let us take $c_1 = y_1^0, \cdots, c_n = y_n^0$ where y_i^0 stands for the value of y_i at $t = t_0$. Then it has been established that the general solution

$$y_i = F_i(t, c_1, \cdots, c_n) \qquad (i = 1, \cdots, n)$$

is analytic in $c_1, \cdots, c_n$ for $|c_i|$, $(i = 1, \cdots, n)$, small and for any t lying in the t interval for $c_1 = \cdots = c_n = 0$. But the solution is then $x_i = 0$ $(i = 1, \cdots, n)$ for *all* values of t, so that the solution is analytic in $c_1, \cdots, c_n$, t for t arbitrarily large, provided that $|c_i|$ are then sufficiently small. Thus the right-hand members may be expanded in power series in $c_1, \cdots, c_n$ with coefficients analytic in t for all values of t. Furthermore we have clearly for $t = t_0$

$$\partial \varphi_i / \partial c_j = \delta_{ij} \quad (\delta_{ii} = 1;\ \delta_{ij} = 0,\ i \neq j).$$

Hence $\partial \varphi_i / \partial c_j$ $(i = 1, \cdots, n)$ constitute n linearly independent solutions of the equations of variation for $j = 1, \cdots, n$, and the determinant $|\partial \varphi_i / \partial c_j|$ will not vanish for any t.

There exist formal solutions of any system (3).

It is obvious that if conjugate variables are introduced in the fashion of the preceding section, then the formal solutions will have a corresponding modified form.

The significance of the formal solutions will be specified further in the following chapter. Suffice it to say here that these are the type of solutions actually employed in astronomical problems when the perturbations of a periodic motion need to be calculated.

4. The equilibrium problem. As has been pointed out earlier (section 1), the simplest case for consideration is that presented by an ordinary equilibrium point. For this case the functions $X_1, \cdots, X_n$ do not involve t explicitly. In dealing with this special case we shall use only transformations of the formal group which do not involve the time t.

The equations of variation take the form

$$dy_i/dt = \sum_{j=1}^{n} c_{ij} y_j \qquad (i = 1, \cdots, n)$$

where the constants c_{ij} are the values of $\partial X_i / \partial x_j$ evaluated at the origin. In the case to which we limit attention at first the determinant equation

$$|c_{ij} - m\,\delta_{ij}| = 0 \tag{6}$$

has n roots $m_1, \cdots, m_n$, not subject to any commensurability relation

$$i_1 m_1 + \cdots + i_n m_n = 0 \tag{7}$$

for any set of integers $i_1, \cdots, i_n$ not all zero. These roots are real except possibly for certain conjugate imaginary pairs. It is to be noted that the assumed inequality excludes the possibility of a zero root so that the determinant $|c_{ij}|$ is not zero.

Now we can clearly determine a square array $l_{ij}, (i, j = 1, \cdots, n)$, with l_i not all zero for any particular j, such that the sets $l_{1k}, \cdots, l_{nk}$ satisfy the n homogeneous linear equations

$$\sum_{j=1}^{n} c_{ij}\, l_{jk} = l_{ik}\, m_k. \tag{8}$$

In fact the vanishing of the determinant of this system gives precisely the defining equation for the roots m_k. Moreover the determinant $|l_{ij}|$ is not zero. This fact and the reasoning which follows are of course well known in the ordinary theory of linear transformations; however, for the sake of completeness the reasoning is included. In fact, suppose the contrary to be true, so that multipliers $\varrho_1, \cdots, \varrho_n$ not all zero exist such that

$$\sum_{j=1}^{n} l_{ij}\,\varrho_j = 0 \qquad (i = 1, \cdots, n).$$

Multiplying the earlier equations by ϱ_k, $(k = 1, \cdots, n)$, and adding, there results at once

$$0 = \sum_{k=1}^{n} l_{ik}\, m_k\, \varrho_k$$

so that $m_i\,\varrho_i$, $(i = 1, \cdots, n)$, also yield a second set of such multipliers. Continuing in this way we infer that $m_i\varrho_i$, $m_i\varrho_i^2, \cdots$ must form further sets of multipliers. Hence by linear combination still more general sets of multipliers

$$(c_0 + c_1 m_i + c_2 m_i^2 + \cdots + c_{n-1} m_i^{n-1})\,\varrho_i \quad (i = 1, \cdots, n),$$

are obtained. But the n quantities in parentheses here can be made to take on n arbitrarily assigned values, just because the roots m_i are distinct. In particular the coefficient of any $\varrho_k \neq 0$ can be replaced by 1 while all the others are made to vanish. But this would necessitate that $l_{ik} = 0$, $(i = 1, \cdots, n)$, contrary to hypothesis. Consequently $|l_{ij}|$ cannot vanish.

It is also obvious that we may take the quantities l_{ij} so that in the transformation

$$y_i = \sum_{j=1}^{n} l_{ij}\, z_j \qquad (i = 1, \cdots, n),$$

from y_i to z_i, the variables z_i and z_j corresponding to conjugate imaginary roots m_i and m_j will have conjugate imaginary values when $y_1, \cdots, y_n$ are real, and conversely.

Now let us introduce this change of variables in the special type of equations (1) under consideration. The equations obtained by substitution in (1) of the linear expressions in z_i for y_i will then assume the form

$$\sum_{j=1}^{n} l_{ij} \frac{dz_j}{dt} = \sum_{j=1}^{n} l_{ij} m_j z_j + \cdots \qquad (i = 1, \cdots, n),$$

where the terms of higher than the first degree in $z_1, \cdots, z_n$ in the right-hand members are only indicated; in thus writing the first terms on the right, the characteristic property of the quantities l_{ij} is of course employed. It follows from these relations that the equations in z_i have the form

$$dz_i/dt = m_i z_i + \cdots \qquad (i = 1, \cdots, n)$$

where only the linear terms are explicitly written.

Thus we may take the system to be of the prepared form

$$dx_i/dt = m_i x_i + F_i(x_1, \cdots, x_n) \qquad (i = 1, \cdots, n),$$

in which we may write

$$F_i = F_{i2} + F_{i3} + \cdots \qquad (i = 1, \cdots, n),$$

with F_{ik} a homogeneous polynomial of degree k in $x_1, \cdots, x_n$.

We shall show next that we can obtain formal series

$$\varphi_i(x_1, \cdots, x_n) = \varphi_{i2} + \varphi_{i3} + \cdots \qquad (i = 1, \cdots, n),$$

such that the transformation

$$\bar{x}_i = x_i + \varphi_i \qquad (i = 1, \cdots, n)$$

reduces the differential equations to the form

$$d\bar{x}_i/dt = m_i \bar{x}_i \qquad (i = 1, \cdots, n).$$

This will be achieved, provided that the equations

$$dx_i/dt + d\varphi_i/dt = m_i(x_i + \varphi_i) \qquad (i = 1, \cdots, n)$$

follow in consequence of the differential equations in x_i. Using these equations to eliminate dx_i/dt, we obtain as the desired relations

$$F_i + \sum_{j=1}^{n} (\partial \varphi_i / \partial x_j)(m_j x_j + F_j) = m_i \varphi_i \qquad (i = 1, \cdots, n).$$

On expanding F_i and φ_i in series, these take the form for $i = 1, \cdots, n$,

$$F_{i2} + \sum_{j=1}^{n} \frac{\partial \varphi_{i2}}{\partial x_j} m_j x_j = m_i \varphi_{i2},$$

$$\cdots\cdots\cdots\cdots$$

$$F_{ik} + \sum_{j=1}^{n} \left(\frac{\partial \varphi_{ik}}{\partial x_j} m_j x_j + \sum_{p+q=k} \frac{\partial \varphi_{ip}}{\partial x_j} F_{jq} \right) = m_i \varphi_{ik},$$

$$\cdots\cdots\cdots\cdots$$

Let us consider the first equation written for any i, which obviously constitutes a partial differential equation for φ_{i2}. The coefficient c_i of the term

$$c_i x_1^{l_1} \cdots x_n^{l_n} \qquad (l_1 + \cdots l_n = 2)$$

in φ_{i2} is then evidently determined in terms of the analogous coefficient d_i of F_{i2} be means of the n equations

$$d_i + [l_1 m_1 + \cdots + (l_i - 1) m_i + \cdots + l_n m_n] c_i = 0.$$

But the term in parenthesis is not zero by virtue of the hypothesis made concerning the quantities m_i, so that c_i can be determined as desired. Hence there is a unique set of homogeneous quadratic polynomials φ_{i2} satisfying the first set of the equations written above.

In the same way the second set of equations determines φ_{i3} uniquely since the equation for determination of the coefficients in φ_{i3} is of the same general type as above except that we have $l_1 + \cdots + l_n = 3$ in this case.

Ignoring then the questions of convergence of the series employed we arrive at the following conclusion:

By means of a formal transformation

$$x_i = f_i(z_1, \cdots, z_n) = \sum_{j=1}^{n} l_{ij} z_j + \frac{1}{2} \sum_{j,k=1}^{n} l_{ijk} z_j z_k + \cdots \qquad (i = 1, \cdots, n)$$

with $|l_{ij}| \neq 0$, *the differential equations* (1) *with an ordinary equilibrium point of general type at the origin, can be reduced to the normal form*

$$dz_i/dt = m_i z_i \qquad (i = 1, \cdots, n), \tag{9}$$

so that corresponding to conjugate roots m_i *and* m_j *there are conjugate variables* z_i *and* z_j.

Since the normal form just written is integrable with general solution

$$z_i = c_i e^{m_i t} \qquad (i = 1, \cdots, n)$$

we may state the following conclusion also:

The corresponding formal solution of (1) *may be written in the form*

$$x_i = f_i(c_1 e^{m_1 t}, \cdots, c_n e^{m_n t}) \qquad (i = 1, \cdots, n)$$

where the f_i *are the same formal power series as appear in the transformation to normal form.*

5. The generalized equilibrium problem. It is not difficult to extend the above method to the generalized equilibrium problem in which we start with equations of the form (3). Here the equations of variation form a system of n ordinary linear differential equations

$$dy_i/dt = \sum_{j=1}^{n} \frac{\partial X_i}{\partial x_j}\bigg|_{t=0} y_j \qquad (i = 1, \cdots, n)$$

with coefficients $\partial X_i/\partial x_j\big|_{t=0}$ which are analytic periodic functions of t of period τ. Let $y_{1k}, \cdots, y_{nk}$, $(k = 1, \cdots, n)$, give for each k a solution such that the n solutions are linearly independent. Then the general solution is a linear combination of these particular solutions. When t is increased by τ, the equations of variation are unaltered. Hence we have

$$y_{ik}(t+\tau) = \sum_{l=1}^{n} y_{il}(t)\, c_{kl} \qquad (i, k = 1, \cdots, n).$$

Now if we define $m_1, \cdots, m_n$ as in (8) by means of the square array c_{ij} thus obtained, we may choose another linearly independent set of n solutions which will reduce the above relations to a normal form

$$y_{ik}(t+\tau) = m_k\, y_{ik} \qquad (i, k = 1, \cdots, n).$$

We confine ourselves as before to the general case, in that we exclude linear commensurability relations between $\mu_1, \cdots, \mu_m$ and $2\pi\sqrt{-1}/\tau$, where $\mu_k = (\log m_k)/\tau$. In this event $m_1, \cdots, m_n$ are all distinct.

Now let us write the elements appearing in the solution of the equations of variation in the form

$$y_{ik} = e^{\mu_k t} p_{ik}(t) \qquad (i, k = 1, \cdots, n),$$

when it is apparent that the functions p_{ik} will be periodic of period τ. Furthermore from a familiar theorem we know that the determinant

$$|\, y_{ij} \,| = |\, p_{ij} \,|\, e^{(\mu_1 + \cdots + \mu_n)t}$$

is nowhere 0. Consequently the particular linear change of variables

$$x_i = \sum_{j=1}^{n} p_{ij} z_j \qquad (i = 1, \cdots, n)$$

from $x_1, \cdots, x_n$ to $z_1, \cdots, z_n$ is within the admitted group. The equations of variation will have a solution

$$y_i = \delta_{ik} e^{\mu_k t} \qquad (i = 1, \cdots, n)$$

for $k = 1, \cdots, n$, so that the new equations must be

$$dz_i/dt = \mu_i z_i + \cdots \qquad (i = 1, \cdots, n).$$

Consequently we infer that it is no restriction to write the given equations in the prepared form

$$dx_i/dt = \mu_i x_i + F_i(x_1, \cdots, x_n, t) \qquad (i = 1, \cdots, n)$$

where F_i is periodic in t of period τ of course. It is obvious that if certain pairs of the quantities m_i are conjugate imaginaries the transformation of variables employed may be taken to be of the admitted conjugate type.

Suppose that we continue by effecting a further transformation of the same type as in the ordinary equilibrium problem save that the coefficients in the series φ_i need not be constants but may be periodic analytic functions of t of period τ. If we endeavor to choose this set of functions φ_i so as to normalize the transformed differential equations as in the special case of ordinary equilibrium we obtain analogous equations, namely

$$F_{i2} + \frac{\partial \varphi_{i2}}{\partial t} + \sum_{j=1}^{n} \frac{\partial \varphi_{i2}}{\partial x_j} \mu_j x_j = \mu_i \varphi_{i2},$$

$$\cdots\cdots\cdots\cdots\cdots\cdots\cdots\cdots\cdots$$

$$F_{ik} + \frac{\partial \varphi_{ik}}{\partial t} + \sum_{j=1}^{n} \left(\frac{\partial \varphi_{ik}}{\partial x_j} \mu_j x_j + \sum_{p+q=k} \frac{\partial \varphi_{ip}}{\partial x_j} F_{jq} \right) = \mu_i \varphi_{ik},$$

$$\cdots\cdots\cdots\cdots\cdots\cdots\cdots\cdots\cdots$$

On considering a typical term in φ_{i2},

$$c_i(t) x_1^{l_1} \cdots x_n^{l_n} \qquad (l_1 + \cdots + l_n = 2),$$

we find as the required conditions

$$d_i(t) + \frac{dc_i}{dt} + \left[l_1 \mu_1 + \cdots + (l_i - 1) \mu_i + \cdots + \mu_n \right] c_i = 0 \qquad (i = 1, \cdots, n)$$

where d_i is the like coefficient in F_{i2}. Here the coefficient λ of c_i is not zero, and it is immediately possible to solve for c_i,

$$c_i(t) = k_i e^{-\lambda t} - e^{-\lambda t} \int_0 d(t) e^{\lambda t} \, dt.$$

This solution will be periodic in t of period τ if and only if

$$k_i (1 - e^{\lambda \tau}) = \int_0^{\tau} d_i(t) e^{\lambda t} \, dt.$$

It is possible to determine k_i in one and only one way so that this equation holds, provided that λ is not an integral multiple of $2\pi\sqrt{-1}/\tau$. But this relation would require a commensurability relation between the 'multipliers' $\mu_1, \cdots, \mu_n$ and $2\pi\sqrt{-1}/\tau$ of the excluded type.

Thus, as before, no difficulty arises in determining $\varphi_{i2}, \varphi_{i3}, \cdots$, in succession in such fashion that the desired normal form is obtained.

By means of a formal transformation

$$x_i = f_i(z_1, \cdots, z_n, t) = \sum_{j=1}^{n} l_{ij}(t) z_j + \frac{1}{2} \sum_{j,k=1}^{n} l_{ijk}(t) z_j z_k + \cdots \qquad (i = 1, \cdots, n)$$

with l_{ij} analytic in t and periodic of period τ, and $|l_{ij}(t)| \neq 0$, the differential equations (3) *with a generalized equilibrium point at the origin of general type can also be reduced to the normal form*

$$dz_i/dt = \mu_i z_i \qquad (i = 1, \cdots, n). \tag{10}$$

The corresponding formal solution of (3) *is then evidently*

$$y_i = f_i(c_1 e^{\mu_1 t}, \cdots, c_n e^{\mu_n t}, t) \qquad (i = 1, \cdots, n).$$

6. On the Hamiltonian multipliers. As a first step toward obtaining an analogous normal form for a Hamiltonian system of equations at an equilibrium point, we demonstrate some fundamental well-known properties of the multipliers in this case.*

Here the equations occur in the particular form

$$\frac{dp_i}{dt} = -\frac{\partial H}{\partial q_i}, \quad \frac{dq_i}{dt} = \frac{\partial H}{\partial p_i} \qquad (i = 1, \cdots, m) \tag{11}$$

where H is a real analytic function of $n = 2m$ variables $p_1, \cdots, q_m$. If these equations have an equilibrium point at

* Cf. Poincaré, *Les Méthodes nouvelles de la Mécanique céleste*, vol. 1, chap. 4. His 'characteristic exponents' are our multipliers.

the origin, then evidently all of the first partial derivatives of H vanish at the origin and if we ignore an additive constant in H we may write

$$H = H_2 + H_3 + \cdots,$$

where H_k is a homogeneous polynomial of degree k in these dependent variables and where in particular we have

$$H_2 = \sum_{j,k=1}^{m} \left(\frac{1}{2} a_{jk} p_j p_k + b_{jk} p_j q_k + \frac{1}{2} c_{jk} q_j q_k \right).$$

Here we may take $a_{ij} = a_{ji}$, $c_{ij} = c_{ji}$, but b_{ij} are in general distinct from b_{ji}.

The equations of variation are obtained by replacing H by H_2 and p_i, q_i by P_i, Q_i, and may be written in the explicit form

$$\begin{aligned} dP_i/dt &= -\sum_{j=1}^{m} b_{ji} P_j - \sum_{j=1}^{m} c_{ij} Q_j, \\ dQ_i/dt &= \sum_{j=1}^{m} a_{ij} P_j + \sum_{j=1}^{m} b_{ij} Q_j \quad (i = 1, \cdots, m) \end{aligned}$$

which is a particular type of system of $2m$ linear differential equations of the first order with constant coefficients.

Our first remark is merely to the effect that *in general* the multipliers will be distinct. To verify this fact, it is merely necessary to exhibit the $2m$ exponential solutions in a single special case. If we take

$$H_2 = \sum_{j=1}^{m} \frac{1}{2} \mu_j (p_j^2 + q_j^2),$$

the equations of variation reduce to

$$dP_i/dt = -\mu_i Q_i, \quad dQ_i/dt = \mu_i P_i \quad (i = 1, \cdots, m)$$

with $2m$ particular solutions

$$P_i = \delta_{ik} e^{\pm \mu_k \sqrt{-1}\, t}, \qquad Q_i = \mp \delta_{ik} e^{\pm \mu_k \sqrt{-1}\, t}$$

where δ_{ij} has its usual significance. Hence the $2m$ values of the multipliers are $\pm\mu_k\sqrt{-1}$, $(k=1,\cdots,m)$, and these will be distinct if $\mu_1,\cdots,\mu_m$ are distinct positive numbers, for instance.

Before leaving the special case just cited we note that if H_2 is of this special form, we can introduce conjugate variables

$$\xi_i = p_i + \sqrt{-1}\,q_i, \quad \eta_i = p_i - \sqrt{-1}\,q_i \quad (i=1,\cdots,m),$$

and then we find

$$\frac{d\xi_i}{dt} = -\frac{\partial \bar{H}}{\partial \eta_i}, \quad \frac{d\eta_i}{dt} = \frac{\partial \bar{H}}{\partial \xi_i} \quad (i=1,\cdots,m)$$

where $\bar{H} = -2\sqrt{-1}\,H$ and $\bar{H}_2$ takes the normal form

$$\bar{H}_2 = -\sum_{j=1}^{m} \mu_j \sqrt{-1}\,\xi_j\,\eta_j.$$

Consequently under this type of change of variables the Hamiltonian form of the differential equations is maintained. In this event the equations of variation are still simpler, namely of the form

$$d\xi_i/dt = \mu_i\sqrt{-1}\,\xi_i, \quad d\eta_i/dt = -\mu_i\sqrt{-1}\,\eta_i \quad (i=1,\cdots,m).$$

This type of conjugate variables plays an important role later on.

Let us suppose then that we are confronted by the general case in which the $2m$ multipliers are distinct. We propose to show that these quantities occur in m distinct pairs, each one the negative of the multiplier paired with it. This has already been seen above to be true in the special case cited.

Since the multipliers are distinct by hypothesis, a complete set of solutions

$$P_{1k},\cdots,P_{mk},\ Q_{1k},\cdots,Q_{mk} \quad (k=1,\cdots,2m),$$

exists of the form

$$P_{ik} = C_{ik}\,e^{\lambda_k t},\ Q_{ik} = D_{ik}\,e^{\lambda_k t} \quad (i=1,\cdots,m,\ k=1,\cdots,2m),$$

where the corresponding determinant of constants of order $2m$ formed from C_{ik}, D_{ik} is not 0. Such a complete set of particular solutions has the property that the most general solution is expressible as a linear combination of these particular ones.

But if P_i, Q_i and P_i^*, Q_i^* are any two solutions of the equations of variation we have

$$\sum_{j=1}^{m} (Q_j P_j^* - P_j Q_j^*) = \text{const.}$$

This fact is readily verified by differentiation with respect to t, and use of the equations of variation, when it is seen that the derivative of the left-hand side reduces identically to zero.

If then we substitute in this integral relation, pairs of the above particular solutions we find

$$\sum_{j=1}^{m} (D_{ik} C_{il} - C_{ik} D_{il}) e^{(\lambda_k + \lambda_l)t} = \text{const.}$$

for all k and l. This clearly implies at once that either $\lambda_k + \lambda_l$ is 0 or that the constant on the right-hand side is 0. A proper use of this fact will lead us easily to the desired conclusion.

If each λ_i has a corresponding λ_j such that $\lambda_i + \lambda_j = 0$, then there is clearly only one such root and the property under consideration is proved. But in the contrary case some root as λ_k has no value so paired with it. Hence, in the integral relations deduced above, the right-hand members must vanish for $l = 1, \cdots, 2m$ if k has this value, whence we find

$$D_{1k} C_{1l} + \cdots + D_{mk} C_{ml} - C_{1k} D_{1l} - \cdots - C_{mk} D_{ml} = 0 \qquad (l = 1, \cdots, 2m).$$

These $2m$ equations are linear and homogeneous in $D_{1k}, \cdots, D_{mk}, -C_{1k}, \cdots, -C_{mk}$, so that the determinant of their coefficients would necessarily vanish. But this determinant is precisely the determinant of order $2m$ referred to above which cannot be zero. Hence there is no such root.

In general at an equilibrium point for the Hamiltonian equations the multipliers can be grouped in m pairs $\lambda_i, -\lambda_i,$ $(i = 1, \cdots, m)$, *and are all distinct.*

It is plain that in general the multipliers are real or are grouped in conjugate imaginary pairs with distinct moduli. Consequently the conjugate of an imaginary multiplier must coincide with its negative. By passing to the special cases by a limiting process we conclude further:

The multipliers λ_i *are either real or pure imaginary quantities.*

We define the general equilibrium point of this type as the one in which $\lambda_1, \cdots, \lambda_m$ are not subject to any linear commensurability relation of the type (7), and confine attention to this general case.

7. Normalization of H_2. Assuming then that the equilibrium point of the Hamiltonian system under consideration is of this general type, we can effect a linear transformation of variables

$$p_i = \sum_{j=1}^{m} (d_{ij}\bar{p}_j + e_{ij}\bar{q}_j), \qquad q_i = \sum_{j=1}^{m} (f_{ij}\bar{p}_j + g_{ij}\bar{q}_j) \qquad (i = 1, \cdots, m),$$

which reduces the corresponding equations of variation to the normal form

$$dP_i/dt = \lambda_i P_i, \qquad dQ_i/dt = -\lambda_i Q_i \qquad (i = 1, \cdots, m).$$

In fact this reduction (see section 4) merely required that the roots of the characteristic equation (6) be distinct, as is here the case. Of course the associated pairs $\bar{p}_i, \bar{q}_i$ are taken as corresponding to associated roots $\lambda_i, -\lambda_i$. If λ_i is real, $\bar{p}_i, \bar{q}_i$ are real variables. If λ_i is a pure imaginary, $\bar{p}_i, \bar{q}_i$ are conjugate variables.

We propose to demonstrate that this linear transformation does not destroy the Hamiltonian form of the equations.

To begin with we observe that the equations of variation can be written in the variational Hamiltonian form

$$\delta \int_{t_0}^{t_1} \left[\sum_{j=1}^{m} P_i Q_i' - H_2 (P_1, \cdots, Q_m) \right] dt = 0,$$

in which H is replaced by its second degree terms. Under the above linear change of variables this evidently takes the form

$$\delta \int_{t_0}^{t_1} \Big[\sum_{j,k=1}^{m} (K_{jk} P_j P_k' + L_{jk} Q_j P_k' + M_{jk} P_j Q_k' + N_{jk} Q_j Q_k') - H_2 \Big] dt = 0,$$

where we may write

$$H_2 = \sum_{j,k=1}^{m} (R_{jk} P_j P_k + S_{jk} P_j Q_k + T_{jk} Q_j Q_k).$$

Here the dashes over the letters have been omitted, and we may obviously assume

$$R_{ij} = R_{ji}, \qquad T_{ij} = T_{ji} \qquad (i, j = 1, \cdots, m).$$

Applying the ordinary Lagrangian rule this gives the equations of variation in the new variables,

$$\frac{d}{dt}\Big[\sum_{j=1}^{m}(K_{ji} P_j + L_{ji} Q_j)\Big] - \sum_{j=1}^{m}(K_{ij} P_j' + M_{ij} Q_j') + \sum_{j=1}^{m}(2 R_{ij} P_j + S_{ij} Q_j) = 0,$$
$$(i = 1, \cdots, m),$$

$$\frac{d}{dt}\Big[\sum_{j=1}^{m}(M_{ji} P_j + N_{ji} Q_j)\Big] - \sum_{j=1}^{m}(L_{ij} P_j' + N_{ij} Q_j') + \sum_{j=1}^{m}(S_{ji} P_j + 2 T_{ij} Q_j) = 0,$$
$$(i = 1, \cdots, m).$$

But the solutions of these equations are known. In particular we have a solution

$$P_i = \delta_{ik} e^{\lambda_k t}, \qquad Q_i = 0 \qquad (i = 1, \cdots, m)$$

which when substituted in the first of the above equations gives at once

$$\lambda_k (K_{ki} - K_{ik}) + 2 R_{ik} = 0.$$

Interchanging i and k, and noting that $R_{ik} = R_{ki}$ we infer further for any i and k

$$(\lambda_i - \lambda_k)(K_{ki} - K_{ik}) = 0,$$

so that $K_{ki} = K_{ik}$ for i distinct from k as well as for $i = k$. It follows also that R_{ik} vanishes for all i and k.

Similarly we can infer that $N_{ki} = N_{ik}$ and that T_{ik} vanishes from the second set of equations. Thus the terms

$$\sum_{j,k=1}^{m} K_{jk} P_j P'_k, \qquad \sum_{,k=1}^{m} M_{jk} Q_j Q'_k$$

are exact derivatives, and may be omitted under the integral sign. The equations of variation are thus of the more special form

$$\frac{d}{dt}\left[\sum_{j=1}^{m} L_{ji} Q_j\right] - \sum_{j=1}^{m} M_{ij} Q'_j + \sum_{j=1}^{m} S_{ij} Q_j = 0 \qquad (i = 1, \cdots, m),$$

$$\frac{d}{dt}\left[\sum_{j=1}^{m} M_{ji} P_j\right] - \sum_{j=1}^{m} L_{ij} P'_j + \sum_{j=1}^{m} S_{ji} P_j = 0 \qquad (i = 1, \cdots, m).$$

To determine these equations still more completely we substitute

$$P_i = 0, \qquad Q_i = \delta_{ik} e^{-\lambda_k t} \qquad (i = 1, \cdots, m)$$

in the first set of these equations, and obtain immediately for all i and k

$$\lambda_k (M_{ik} - L_{ki}) + S_{ik} = 0.$$

Similarly from the second set we obtain for all i and k,

$$\lambda_k (M_{ki} - L_{ik}) + S_{ki} = 0.$$

Interchanging i and k in this equation, and comparing the equation obtained with the preceding one we infer that for $i \neq k$ we have

$$M_{ik} = L_{ki}, \qquad S_{ik} = 0.$$

In consequence the sum

$$\sum_{j,k=1}^{m} (L_{jk} Q_j P'_k + M_{jk} P_j Q'_k)$$

differs from the sum

$$\sum_{j=1}^{m} L_{jj}\, Q_j\, P_j' + M_{jj}\, P_j\, Q_j',$$

and so from

$$\sum_{j=1}^{m} (M_{jj} - L_{jj})\, P_j\, Q_j'$$

by an exact differential. Thus it is legitimate to write the principle of variation in the specific form

$$\delta \int_{t_0}^{t_1} \left[\sum_{j=1}^{m} (M_{jj} - L_{jj})\, P_j\, Q_j' + \sum_{j=1}^{m} S_{jj}\, P_j\, Q_j \right] dt = 0$$

in such new variables with equations of variation*

$$(M_{ii} - L_{ii})\, Q_j' + S_{ii}\, Q_i = 0, \qquad (M_{ii} - L_{ii})\, P_i' - S_{ii}\, P_i = 0$$
$$(i = 1, \cdots, m),$$

so that we have necessarily

$$(M_{ii} - L_{ii})\, \lambda_i = S_{ii} \qquad (i = 1, \cdots, m).$$

Consider now the simplest case when every root λ_i is real. In this case if we replace the real variable P_i by

$$\bar{P}_i = (M_{ii} - L_{ii})\, P_i \qquad (i = 1, \cdots, m)$$

the variational principle is seen to have the form

$$\delta \int_{t_0}^{t_1} \left[\sum_{j=1}^{m} P_i\, Q_i' - \sum_{j=1}^{m} \lambda_j\, P_j\, Q_j \right] dt = 0.$$

This further change of variables is legitimate inasmuch as p_i, q_i were not determined up to real multipliers in this case. It appears then that the term

$$\sum_{j=1}^{m} P_j\, Q_j'$$

* The constants $M_{ii} - L_{ii}$ are not zero since the equations of variation do not degenerate.

remains of essentially the same form after the linear transformation.

Another case is that in which every λ_i is a pure imaginary quantity. Here by taking the pairs p_i, q_i appropriately we can clearly write the pure imaginary quantities $M_{ii} - L_{ii}$ as $\varrho_i \sqrt{-1}$, $\varrho_i > 0$. Here we may replace P_i, Q_i by

$$\overline{P}_i = \sqrt{\varrho_i}\, P_i, \qquad \overline{Q}_i = \sqrt{\varrho_i}\, Q_i \qquad (i = 1, \cdots, m)$$

when a like variational form is obtained.

It is apparent that this same linear change of variables must preserve the original Hamiltonian form since

$$\sum_{j=1}^{m} p_j\, q_j'$$

is essentially unaltered by this transformation.

By a suitable preliminary linear transformation with constant coefficients any Hamiltonian system with equilibrium point of general type at the origin may be taken in a normalized form in which

$$H_2 = \sum_{j=1}^{m} \lambda_j\, p_j\, q_j.$$

8. The Hamiltonian equilibrium problem. In order to further normalize the Hamiltonian equations in the vicinity of an equilibrium point, we propose to apply a series of transformations

$$p_i = \partial K / \partial q_i, \qquad \overline{q}_i = \partial K / \partial \overline{p}_i \qquad (i = 1, \cdots, m)$$

with

$$K = \sum_{j=1}^{m} \overline{p}_j\, q_j + K_3 + K_4 + \cdots$$

where $K_3, K_4, \cdots$ are homogeneous functions of $\overline{p}_i, q_i$, $(i = 1, \cdots, m)$ of degree indicated by the subscripts. Such transformations have been seen to leave the Hamiltonian form undisturbed and to form a group. It will be observed that if $K_s = 0$ for $s > 2$, the transformation is the identity.

We begin by taking $K_s = 0$, $s > 3$, and by attempting to select K_3 so as to simplify H_3 as far as possible. Here we have

$$p_i = \bar{p}_i + \partial K_3/\partial q_i, \quad \bar{q}_i = q_i + \partial K_3/\partial p_i \quad (i = 1, \cdots, m).$$

If we solve explicitly for $p_i, q_i, (i = 1, \cdots, m)$ in terms of $\bar{p}_i, \bar{q}_i, (i = 1, \cdots, m)$, we will clearly obtain

$$p_i = \bar{p}_i + \partial K_3^*/\partial \bar{q}_i + \cdots, \quad q_i = \bar{q}_i - \partial K_3^*/\partial \bar{p}_i + \cdots \quad (i = 1, \cdots, m),$$

where K_3^* denotes the function obtained by replacing q_i by $\bar{q}_i$ in K_3. The terms explicitly written give the series expansion up to the terms of the third degree. The modified value of H, obtained by direct substitution is

$$H_2(\bar{p}_1 + \partial K_3^*/\partial \bar{q}_1 + \cdots, \cdots \bar{q}_m - \partial K_3^*/\partial \bar{p}_m + \cdots) + H_3 + \cdots,$$

where the arguments of $H_3, H_4, \cdots$ are the same as those of H_2. To terms of the third degree inclusive we find then

$$\bar{H} = \sum_{j=1}^{m} \lambda_j \bar{p}_j \bar{q}_j + \sum_{j=1}^{m} \lambda_j \left(\bar{q}_j \frac{\partial K_3^*}{\partial \bar{q}_j} - \bar{p}_j \frac{\partial K_3^*}{\partial \bar{p}_j} \right) + H_3(\bar{p}_1, \cdots, \bar{q}_m) + \cdots.$$

Thus, as would be expected, the form of H_2 is unmodified while H_3 takes the form

$$\sum_{j=1}^{m} \lambda_j \left(q_j \frac{\partial K_3}{\partial q_j} - p_j \frac{\partial K_3}{\partial p_j} \right) + H_3(p_1, \cdots, q_m)$$

in which K_3 is at our disposal. Now any term in K_3 may be written

$$c p_1^{\alpha_1} \cdots p_m^{\alpha_m} q_1^{\beta_1} \cdots q_m^{\beta_m} \quad (\alpha_1 + \cdots + \beta_m = 3).$$

The corresponding term in the modified H_3 has a coefficient

$$c[\lambda_1(\beta_1 - \alpha_1) + \cdots + \lambda_m(\beta_m - \alpha_m)] + h$$

in which h is the coefficient analogous to c in the original H_3. Moreover the coefficient of c cannot vanish unless

$$\beta_1 = \alpha_1, \cdots, \beta_m = \alpha_m,$$

which is clearly not possible inasmuch as the sum of all the α_i's and β_i's is 2. Thus by proper choice of each c we can make the new H_3 vanish.

If now we proceed to try to eliminate H_4 as far as possible by a further transformation of the same type in which $K_s = 0$ except for $s = 4$ we obtain a transformation

$$p_i = \bar{p}_i + \partial K_4^* / \partial \bar{q}_i + \cdots, \qquad q_i = \bar{q}_i - \partial K_4^* / \partial \bar{p}_i + \cdots \qquad (i = 1, \cdots, m)$$

which does not affect H_2 or $H_3 = 0$, but alters H_4 to

$$\sum_{j=1}^{m} \lambda_j \left(q_j \frac{\partial K_4}{\partial q_j} - p_j \frac{\partial K_4}{\partial p_j} \right) + H_4(p_1, \cdots, q_m).$$

Here we can eliminate the terms of H_4 save those which contain each p_i, q_i to the same degree, namely those of the forms

$$c(p_i q_i)^2, \qquad d p_i q_i p_j q_j \qquad (i, j = 1, \cdots, m),$$

by the same method. For we have

$$\alpha_1 + \cdots + \beta_m = 4$$

in this case, and all terms can be made to disappear except those for which $\alpha_i = \beta_i$, $(i = 1, \cdots, m)$, i. e., those of the stated type.

Thus it is readily seen that by an infinite series of steps we can eliminate from H all terms except the terms in the m products $p_i q_i$.

By suitable transformations of the above types, a Hamiltonian system with equilibrium point of general type at the origin may be taken into a normal Hamiltonian form in which *only the m products $p_1 q_1, \cdots, p_m q_m$ appear in H while H_2 has the special form*

$$\sum_{j=1}^{m} \lambda_j p_j q_j.$$

It may be noted that the linear transformation employed is also a contact transformation* so that in reality this normal form may be obtained through a single formal contact transformation.

In this normal form the general formal solution is at once obtainable. If we write $\pi_i = p_i q_i$, the normal Hamiltonian equations may be written

$$\frac{dp_i}{dt} = -\frac{\partial H}{\partial \pi_i} p_i, \qquad \frac{dq_i}{dt} = \frac{\partial H}{\partial \pi_i} q_i \quad (i = 1, \cdots, m),$$

whence we find formally

$$p_i q_i = c_i \qquad (i = 1, \cdots, m).$$

Thus the series $\partial H / \partial \pi_i$ reduce to constants and we are led to the following conclusion in a purely formal manner:

The general formal solution of the normal Hamiltonian equations near such an equilibrium point has the form

$$p_i = \alpha_i e^{-\gamma_i t}, \qquad q_i = \beta_i e^{\gamma_i t} \quad (i = 1, \cdots, m),$$

where

$$\gamma_i = \partial H (\alpha_1 \beta_1, \cdots, \alpha_m \beta_m) / \partial \pi_i \quad (i = 1, \cdots, m).$$

In terms of the original variables the corresponding solution may be obtained with the aid of the contact transformation relating the given variables and the normal variables.

9. Generalization of the Hamiltonian problem. If

$$p_i = \varphi_i(t), \qquad q_i = \psi_i(t) \quad (i = 1, \cdots, m)$$

is a periodic motion of period τ, and if we write

$$p_i = \varphi_i + \bar{p}_i, \qquad q_i = \psi_i + \bar{q}_i \quad (i = 1, \cdots, m)$$

* See Whittaker, *Analytical Dynamics*, chap. 16.

then the differential equations for the modified variables are of Hamiltonian form with modified principal function

$$\bar{H} = H + \sum_{j=1}^{m} (\varphi_j' \bar{q}_j - \psi_j' \bar{p}_j)$$

where the accents denote differentiation. Furthermore, this transformation may be written as a contact transformation

$$p_i = \partial K / \partial q_i, \qquad \bar{q}_i = \partial K / \partial \bar{p}_i \quad (i = 1, \cdots, m)$$

with

$$K = \sum_{j=1}^{m} (\bar{p}_j q_j + \varphi_j q_j - \psi_j \bar{p}_j).$$

In these new variables H is a function of $\bar{p}_1, \cdots, \bar{q}_m$ and t, periodic of period τ in the last variable, with $\bar{p}_i = \bar{q}_i = 0$, $(i = 1, \cdots, m)$, a solution corresponding to the given periodic motion. Thus, at least in a formal sense (see chapter IV, section 1), the problem reduces to one of generalized equilibrium.

It will be our aim here to show that a reduction to normal form for such a generalized equilibrium problem can be made which is altogether analogous to that made above in the case of ordinary equilibrium.

We shall merely call attention to the modifications necessary in the argument in dealing with this more general problem.

The first difference to which attention needs to be called is to the obvious fact that in the equations of variation the constants a_{ij}, b_{ij}, c_{ij} are replaced by periodic functions of t with period τ. The second difference is that the constants C_{ij}, D_{ij} which appeared in the solutions of these equations are also such periodic functions.

These modifications do not, however, interfere with the argument made that the multipliers may be grouped in m pairs

$$\lambda_1, \ -\lambda_1, \cdots, \lambda_m, \ -\lambda_m$$

where $\lambda_1, \cdots, \lambda_m$ are real or pure imaginary.

The general equilibrium point may here be appropriately defined as that in which there are no linear commensurability relations between the $m+1$ quantities $\lambda_1, \cdots, \lambda_m$ and $2\pi\sqrt{-1}/\tau$.

In the corresponding linear transformation the coefficients d_{ij}, e_{ij}, f_{ij}, g_{ij} are periodic in t of period τ. Likewise in the variational principle the quantities K_{ij}, L_{ij}, M_{ij}, N_{ij}, R_{ij}, S_{ij}, T_{ij} are similar functions. In determining the form of these functions one finds modified conditions such as that

$$\lambda_k(K_{ki} - K_{ik}) + \frac{dK_{ki}}{dt} + 2R_{ik} = 0.$$

If i and k be interchanged and the results subtracted, there is obtained

$$\frac{d}{dt}(K_{ki} - K_{ik}) + (\lambda_k - \lambda_i)(K_{ki} - K_{ik}) = 0.$$

This differential equation in $(K_{ki} - K_{ik})$ has no periodic solution of period τ (other than 0), just because $\lambda_1, \cdots, \lambda_m$ are of the general type assumed. Hence we infer as before that K_{ik}, K_{ki} are equal while $2R_{ik}$ is $-dK_{ik}/dt$.

But in this case

$$\sum_{j,k=1}^{m} K_{jk}\, p_j\, p'_k$$

is an exact differential if

$$\frac{1}{2}\sum_{j,k=1}^{m} \frac{dK_{jk}}{dt} p_j p_k$$

be added while the negative of this last expression may be incorporated in H. Thus it is clear that we may assume

$$K_{ij} = N_{ij} = R_{ij} = T_{ij} = 0$$

as before. In fact similar slight modifications show that the same normal form for H_2 is obtained by this linear transformation in the generalized equilibrium problem as in the ordinary equilibrium problem.

To make clear that the analogy is complete in dealing with H_3, H_4, $\cdots$, let us consider the new H_3 obtained by a transformation

$$p_i = \bar{p}_i + \partial K_3 / \partial q_i, \qquad \bar{q}_i = q_i + \partial K_3 / \partial \bar{p}_i \quad (i = 1, \cdots, m),$$

where K_3 is homogeneous of the third degree in $\bar{p}_i$, q_i, $(i = 1, \cdots, m)$ with coefficients which are periodic in t of period τ. The new form of H_3 is

$$\frac{\partial K_3^*}{\partial t} + \sum_{j=1}^{m} \lambda_j \left(q_j \frac{\partial K_3^*}{\partial q_j} - p_j \frac{\partial K_3^*}{\partial p_j} \right) + H_3(p_1, \cdots, q_m).$$

The terms

$$c(t) p_1^{\alpha_1} \cdots q_m^{\beta_m}, \qquad h(t) p_1^{\alpha_1} \cdots q_m^{\beta_m} \qquad (\alpha_1 + \cdots + \beta_m = 2)$$

in K_3^* and H_2 respectively lead to a total corresponding equation

$$\frac{dc}{dt} + c[\lambda_1(\beta_1 - \alpha_1) + \cdots + \lambda_m(\beta_m - \alpha_m)] + h = 0,$$

in the attempt to eliminate such a term. This ordinary non-homogeneous linear equation of the first order will have one and only one periodic solution inasmuch as the coefficient of c is incommensurable with $2\pi\sqrt{-1}/\tau$, since we are in the general case. Thus H_3 can be made to disappear.

Likewise all the terms of H_4 can be made to disappear save those in the products $p_i q_i$, $(i = 1, \cdots, m)$. The coefficients in these latter terms can be replaced by constants however; in fact this demand leads to an equation of the form

$$\frac{dc}{dt} + h(t) = C$$

where C is an arbitrary constant at our disposal; thus we find

$$c = \int (C - h(t))\, dt,$$

which is obviously periodic of period τ if C is chosen as the mean value of $h(t)$ over a period. Hence we are led to the same conclusion as before.

By means of such a series of transformations of the generalized Hamiltonian equilibrium problem, the Hamiltonian function *may be given the same normal form as was obtained in the case of ordinary equilibrium.**

10. On the Pfaffian multipliers. Suppose now that we take an extended Pfaffian variation problem

$$(12)\quad \delta \int_{t_0}^{t_1} \left[\sum_{j=1}^{2m} X_j(x_1, \cdots, x_{2m}) x_j' + Z(x_1, \cdots, x_{2m}) \right] dt = 0$$

which leads at once to the system of ordinary equations of order $2m$

$$(13)\quad \sum_{j=1}^{2m} \left(\frac{\partial X_i}{\partial x_j} - \frac{\partial X_j}{\partial x_i} \right) \frac{d x_j}{d t} - \frac{\partial Z}{\partial x_i} = 0 \quad (i = 1, \cdots, 2m).$$

We propose to consider these equations in the case when there is an equilibrium point at the origin, under the assumption that the $2m$ analytic functions X_i are such that the skew-symmetric determinant

$$\left| \frac{\partial X_i}{\partial x_j} - \frac{\partial X_j}{\partial x_i} \right|$$

is not 0 at the origin. The constant terms in the series for the functions X_i may obviously be omitted throughout.

It is clear that the Hamiltonian equations appear as a special case of these Pfaffian equations (12).

As will be shown in the following chapter, this generalization of the Hamiltonian equations possesses the same property of automatically fulfilling all of the conditions for complete stability, once the obvious conditions for first order stability are satisfied. Hence from this point of view the Pfaffian equations seem as significant for dynamics as the Hamiltonian equations, although more general in type. Moreover they possess the additional advantage of maintaining their Pfaffian form under an arbitrary transformation of the formal group.

* The results of this chapter were announced in my Chicago Colloquium lectures of 1920. The material so far given is obviously in close relation with previous work, and, in particular, the normal form in the Hamiltonian case is in relation with the formal trigonometric series in dynamics treated for instance in Whittaker, *Analytical Dynamics*, chap. 15.

In fact it is only necessary to substitute the new variables under the integral sign in (12) to obtain the transformed functions X_i and Z.

As a first step in the direction of obtaining a normal form for the Pfaffian equations at an equilibrium point, we propose to prove that for these equations just as for the Hamiltonian equations the multipliers are associated in pairs $\lambda_i, -\lambda_i$.

To begin with we observe that in general these roots must be distinct since they are distinct in the Hamiltonian sub-case.

Now let us make the linear transformation with constant coefficients which takes the equations of variation into normal form. This does not affect the Pfaffian form of course. The corresponding equations of variation obtained from (13) are

$$\sum_{j=1}^{2m}\left[\left(\frac{\partial X_i}{\partial x_j}-\frac{\partial X_j}{\partial x_i}\right)\frac{dy_j}{dt}-\frac{\partial^2 Z}{\partial x_i\,\partial x_j}y_j\right]=0$$

and these must have the particular solutions

$$y_i = \delta_{ik}\,e^{\lambda_k t} \qquad (i = 1,\cdots,2m)$$

for $k = 1,\cdots,2m$. It is understood that the partial derivatives involved in the equations of variation are evaluated at the origin.

Substituting in these particular solutions we obtain readily

$$\left(\frac{\partial X_i}{\partial x_k}-\frac{\partial X_k}{\partial x_i}\right)\lambda_k-\frac{\partial^2 Z}{\partial x_i\,\partial x_k}=0.$$

If we interchange i and k here, and subtract the equation so obtained from the one last written, we find

$$\left(\frac{\partial X_i}{\partial x_k}-\frac{\partial X_k}{\partial x_i}\right)(\lambda_k+\lambda_i)=0 \qquad (i,k=1,\cdots,m).$$

But if for each k we do not have $\lambda_k+\lambda_i=0$ for some i, it would follow from these equations that the skew-symmetric determinant specified above would necessarily vanish. This is impossible since the equations of variation would then degenerate. Hence there is associated necessarily with the root λ a second root $\lambda_i = -\lambda_k$. This is precisely what we desired to prove.

In the Pfaffian equilibrium problem the multipliers also occur in pairs, one of each pair being the negative of the other. These multipliers may be designated by $\lambda_1, -\lambda_1, \cdots, \lambda_m, -\lambda_m$, *and will be real or pure imaginary quantities.*

It is clear that the general case is to be defined as that in which there are no linear relations of commensurability between $\lambda_1, \cdots, \lambda_m$ just as was done in the special Hamiltonian case. We shall restrict ourselves to this general case.

11. Preliminary normalization in Pfaffian problem. It is very easy to establish the fact that the normalization used in the preceding section makes the first degree terms in $X_1, \cdots, X_{2m}$ take essentially the Hamiltonian form. Indeed if we call the $2m$ dependent variables $p_1, \cdots, p_m, q_1, \cdots, q_m$, in such wise that p_i, q_i correspond to paired multipliers $\lambda_i, -\lambda_i$ and if we let $P_1, \cdots, Q_m$ denote the coefficients of $p_1', \cdots, q_m'$ respectively under the integral sign in (12), the previously obtained equations between the partial derivatives $\partial X_i/\partial x_j$ at the origin take the form

$$\frac{\partial P_i}{\partial p_j} = \frac{\partial P_j}{\partial p_i}, \quad \frac{\partial Q_i}{\partial q_j} = \frac{\partial Q_j}{\partial q_i} \qquad (i, j = 1, \cdots, m),$$

$$\frac{\partial P_i}{\partial q_j} = \frac{\partial Q_i}{\partial p_j} \qquad (i, j = 1, \cdots, m;\ i \neq j).$$

The first sets of equations show that the linear terms in P_i involving $p_1, \cdots, p_m$ correspond to an exact differential, as do the linear terms of Q_i in $q_1, \cdots, q_m$. Similarly the second last set of equations shows that the term of P_i involving q_j $(i \neq j)$ together with the corresponding term of Q_j involving p_i combine in a like manner. All of these terms may be omitted, and there remains for consideration only terms

$$\sum_{j=1}^{m} (c_j p_j \, d q_j + d_j q_j \, d p_j),$$

which evidently may be replaced by

$$\sum_{j=1}^{m} (c_j - d_j)\, p_j\, dq_j.$$

In these terms no $c_i - d_i$ can vanish, because of the hypothesis that the fundamental skew-symmetric determinant does not vanish at the origin.

If then p_i, q_i are real variables we may make the further linear transformation

$$\bar{p}_i = p_i, \quad \bar{q}_i = (c_i - d_i)\, q_i \quad (i = 1, \cdots, m)$$

to obtain the desired linear Hamiltonian term. On the other hand if p_i, q_i are conjugate variables then c_i and d_i are conjugate imaginaries and $c_i - d_i$ is a pure imaginary quantity $\varrho\sqrt{-1}$. Here we may set

$$\bar{p}_i = \sqrt{\varrho}\, p_i, \quad \bar{q}_i = \sqrt{\varrho}\, q_i$$

if $\varrho > 0$. If $\varrho < 0$, we may interchange the roles of p_i and q_i.

Let us turn next to consider the function Z. Since we have an equilibrium point at the origin, it is plain that $\partial Z/\partial p_i$, $\partial Z/\partial q_i$, vanish there for $i = 1, 2, \cdots, m$, i. e., that there are no linear terms in Z. The lowest terms in Z are then of the second degree.

Thus it is apparent that the equations of variation, which depend only on these first degree terms in $X_1, \cdots, X_{2m}$, and upon the second degree terms in Z, are of the same type as in the Hamiltonian case. In consequence the same linear transformation employed to obtain a normal form for these lowest degree terms gives $-Z_2$ the form of H_2 in the Hamiltonian case.

We can summarize our results as follows:

By a preliminary linear transformation the Pfaffian equations with an equilibrium point of general type at the origin can be written

$$\delta \int_{t_0}^{t_1} \left[\sum_{j=1}^{m} (P_j q_j' + Q_j p_j') - R \right] dt = 0$$

where

$$P_i = p_i + P_{i2} + \cdots, \quad Q_i = * + Q_{i2} + \cdots \quad (i = 1, \cdots, m),$$

$$R = \sum_{j=1}^{m} \lambda_j p_j q_j + R_3 + \cdots.$$

12. The Pfaffian equilibrium problem. After this preparatory work it is a simple matter to establish the general result, which is that by means of point transformations (independent of t), it is possible to reduce the Pfaffian type of equations to Hamiltonian form. More precisely we propose to show that it is possible to reduce Q_i, $(i = 1, \cdots, m)$ to 0 by a suitable succession of such transformations without interfering with the normal form of P_i. When this has been accomplished, it is merely necessary to write

$$\bar{p}_i = P_i, \quad \bar{q}_i = q_i \qquad (i = 1, \cdots, m)$$

to obtain complete Hamiltonian form in the case when $p_1, \cdots, q_m$ are real variables. A slight modification is necessary in case the variables p_i, q_i are not all real.

In the real case suppose that we write

$$p_i = \bar{p}_i, \quad q_i = \bar{q}_i + G_{i2}(\bar{p}_1, \cdots, \bar{q}_m) \qquad (i = 1, \cdots, m),$$

where, according to our usual notation, G_{i2} is a homogeneous quadratic polynomial in its arguments. The variational principle takes a corresponding form in which the new coefficient P_i has also an initial first degree term p_i as desired, while for the new Q_i we find readily expressions in series

$$* + Q_{i2} + \sum_{j=1}^{m} p_j \frac{\partial G_{j2}}{\partial p_i} + \cdots \qquad (i = 1, \cdots, m).$$

Here the linear terms are lacking, as desired, and only the quadratic terms are written explicitly.

Now we have

$$d\left(\sum_{j=1}^{m} p_j G_{j2}\right) = \sum_{j,k=1}^{m} \left(p_j \frac{\partial G_{j2}}{\partial p_k} + G_{k2}\right) dp_k$$
$$+ \sum_{j,k=1}^{m} \left(p_j \frac{\partial G_{j2}}{\partial q_k} dq_k\right).$$

This identity shows that, by subtracting an exact differential under the integral sign, we may modify the new Q_{i2} to the form $Q_{i2} - G_{i2}$ without introducing any first degree terms in P_i. Hence if we take $G_{i2} = Q_{i2}, (i = 1, \cdots, m)$, the third degree terms in Q_{i2} will have been eliminated.

Next by a further transformation

$$p_i = \bar{p}_i, \qquad q_i = \bar{q}_i + \bar{G}_{i3} \quad (i = 1, \cdots, m)$$

we can similarly eliminate the third degree term in Q_i. Proceding thus indefinitely we arrive at a variational form

$$\delta \int_{t_0}^{t_1} \left[\sum_{j=1}^{m} P_j q_j' + * - R \right] dt = 0$$

where

$$P_i = p_i + P_{i2} + \cdots \qquad (i = 1, \cdots, m),$$

which can evidently be given Hamiltonian form in the manner indicated.

In case some of the pairs of variables p_i, q_i are conjugate imaginaries we may first perform the simple linear transformations to corresponding real variables

$$p_i = (\bar{p}_i + \bar{q}_i \sqrt{-1}\,)/\sqrt{2}, \qquad q_i = (\bar{p}_i - \bar{q}_i \sqrt{-1}\,)/\sqrt{2}$$

so that the term $p_i q_i'$ is replaced by $\bar{p}_i \bar{q}_i'$ except for an exact differential. Thus the normal form for P_i, Q_i is maintained for these real variables. Operating with them as indicated we can reach the same conclusion in this case also.

By a suitable transformation of the formal group, say

$$p_i = \varphi_i(\bar{p}_1, \cdots, \bar{q}_m), \quad q_i = \psi_i(\bar{p}_1, \cdots, \bar{q}_m) \quad (i = 1, \cdots, m),$$

the general Pfaffian equilibrium problem may be made to assume Hamiltionian form.

13. Generalization of the Pfaffian problem. Under the above circumstances it is natural to expect that Pfaffian equations containing the time t, with generalized equilibrium point at the origin, admit of formal reduction to Hamiltonian form. It is not difficult to establish the truth of this conjecture on the basis of certain slight modifications of the above discussion.

In the case of such equilibrium the equations are defined by

$$(14)\quad \delta \int_{t_0}^{t_1} \left[\sum_{j=1}^{2m} X_j(x_1, \cdots, x_{2m}, t)\, x_j' + Z(x_1, \cdots, x_{2m}, t) \right] dt = 0$$

with X_i, $(i = 1, \cdots, 2m)$, and Z periodic in t, that is by

$$(15)\quad \sum_{j=1}^{2m} \left(\frac{\partial X_i}{\partial x_j} - \frac{\partial X_j}{\partial x_i} \right) \frac{dx_j}{dt} + \frac{\partial X_i}{\partial t} - \frac{\partial Z}{\partial x_i} = 0 \qquad (i = 1, \cdots, 2m).$$

In the first place it is obvious that the multipliers are in general distinct by consideration of the same special case as was taken up in the ordinary equilibrium problem.

Furthermore, the equations of variation may again be normalized by a linear transformation in which the coefficients involved are periodic analytic functions of t of period τ, so as to have the solutions

$$y_i = \delta_{ik}\, e^{\lambda_k t} \qquad (i = 1, \cdots, 2m)$$

for $k = 1, \cdots, 2m$. It follows that the multipliers λ_i occur in pairs, each one of a pair being the negative of the other, by essentially the same argument as was used in the equilibrium problem.

Moreover the same argument shows that the linear terms in P_i, Q_i lead to certain exact differentials and terms which may be absorbed in R, so that the same normal form for the first order terms in P_i, Q_i and for the second order terms in R is obtained as before.

Finally as in section 11 we write

$$p_i = \bar{p}_i, \quad q_i = \bar{q}_i + \bar{G}_{i2} \qquad (i = 1, \cdots, m)$$

where now $\bar{G}_{i2}$ has coefficients which are periodic in t of period τ.

Then by an obvious modification of the argument there made, we can make $Q_{i2}=0$ for $i=1,\cdots,m$, and then in succession $Q_{i3}=0,\cdots$.

By a suitable transformation of the formal group the generalized Pfaffian problem of periodic motion may be made to assume Hamiltonian form. Hence the normal form in the Hamiltonian case serves also in the Pfaffian case.

CHAPTER IV

STABILITY OF PERIODIC MOTIONS

1. On the reduction to generalized equilibrium. For motion near equilibrium of a Hamiltonian or, more generally, of a Pfaffian system, the stable case is naturally defined as that in which the multipliers $\lambda_1, \cdots, \lambda_m$ are pure imaginaries, at least provided that there are no linear commensurability relations between these multipliers.

In this chapter, however, we shall limit attention to the analogous but somewhat more complicated question of stability for motion near a periodic motion of such a system.* The method employed involves a reduction to the case of generalized equilibrium. In the more general Pfaffian case this can be accomplished by a change of variables

$$x_i = \bar{x}_i + \varphi_i(t) \qquad (i = 1, \cdots, 2m),$$

in which the periodic functions $\varphi_i(t)$ of period τ are the coördinates of the given periodic motion. By this means the functions $X_1, \cdots, X_{2m}, Z$ are modified (see (12), page 89), since they are no longer independent of t but periodic of period τ; and the given motion now corresponds to generalized equilibrium at the origin in the new $x_1, \cdots, x_{2m}$ space. Hence we are led to consider the question of motion near such a point of generalized equilibrium.

There is, however, a difficulty associated with this reduction to generalized equilibrium which was first signalized by Poincaré for Hamiltonian systems, and which it is desirable to explain briefly.

Following the analogy with the case of ordinary equilibrium, the stable case is defined as that in which the multipliers $\lambda_1, \cdots, \lambda_m$ are pure imaginaries, at least provided that there

* Cf. my article *Stability and the Equations of Dynamics*, Amer. Journ. Math., vol. 49 (1927) for a treatment of the equilibrium problem.

7

are no linear commensurability relations between these multipliers and $2\pi\sqrt{-1}/\tau$. If such relations exist the questions to be considered become more complicated in character.

Unfortunately, for a point of generalized equilibrium obtained by the above method of reduction, the multipliers will not satisfy this condition; more specifically, there will always be a multiplier 0, which is double of course. This may be readily seen. The Pfaffian system admits of the integral $Z =$ const. in the original variables, and therefore admits the integral

$$Z(x_1 + \varphi_1, \cdots, x_{2m} + \varphi_{2m}) = \text{const.}$$

in the modified variables. By differentiation with respect to the $2m$ arbitrary constants in the general solution $x_1, \cdots, x_{2m}$, it appears that the linear relation

$$\frac{\partial Z}{\partial x_1} y_1 + \cdots + \frac{\partial Z}{\partial x_{2m}} y_{2m} = \text{const.}$$

subsists for $2m$ linearly independent solutions $y_1, \cdots, y_{2m}$ of the equations of variation, and so for the most general solution; it is understood that $\partial Z / \partial x_i$, $(i = 1, \cdots, 2m)$, have $\varphi_1, \cdots, \varphi_{2m}$ as arguments. Now if the $2m$ multipliers $\pm\lambda_1, \cdots, \pm\lambda_m$ are distinct, a complete set of $2m$ solutions

$$y_i = p_{ik} e^{\lambda_k t} \qquad (i = 1, \cdots, 2m)$$

for $k = 1, \cdots, 2m$ exists $(\lambda_{m+i} = -\lambda_i)$, in which p_{ij} are of period τ in t. Since $\partial Z / \partial x_i$ are also periodic, substitution of these solutions in the linear integral relations in the y_i's leads immediately to the conclusion that the constants on the right-hand side must vanish, at least for $\lambda_k \neq 0$. But if these constants vanished for such a complete set of solutions, the constants would vanish for every solution $y_1, \cdots, y_{2m}$. This cannot be the case since $y_1, \cdots, y_{2m}$ can be taken arbitrarily for any particular value of t.*

* It is not possible for $\partial Z / \partial x_i$ to vanish simultaneously for $i = 1, \cdots, 2m$ along the original motion, since the Pfaffian equations then yield $dx_i / dt = 0$, $(i = 1, \cdots, 2m)$ which is impossible, the case of ordinary equilibrium being excluded.

There is then a pair of solutions of the equations of variation which belong to the multiplier 0. Now

$$x_i = \varphi_i(t+k) - \varphi_i(t) \quad (i = 1, \cdots, 2m)$$

for any k defines a solution of the given equations after the reduction, so that by differentiation with respect to k, one solution of the equations of variation

$$y_1 = \varphi_1', \cdots, y_{2m} = \varphi_{2m}'$$

is obtained. This has periodic components and so belongs to the multiplier 0. On the other hand the periodic motion with which we start is not isolated, but varies analytically with the constant c in the known integral (i. e., with the energy constant in the Hamiltonian case). This yields the second periodic solution

$$y_1 = \frac{\partial \varphi_1'}{\partial c}, \cdots, y_{2m} = \frac{\partial \varphi_{2m}'}{\partial c},$$

belonging to the multiplier 0. In general there will be no others.

The difficulty may be turned in the following manner. The variable Z may be taken as one of the dependent variables $x_1, \cdots, x_{2m}$, say as x_{2m}, in the original $x_1, \cdots, x_{2m}$ space. Furthermore the variable $\theta = x_{2m-1}$ may be selected as the single angular coördinate, which increases by 2π when a circuit of the curve of periodic motion is made. The remaining coördinates $x_1, \cdots, x_{2m-2}$ may be made to vanish along this curve. Now let us restrict attention to those motions near the given periodic motion for which

$$Z = c$$

has the same value as along this motion. With this understanding, the Pfaffian system becomes of order $2m-1$ in $x_1, \cdots, x_{2m-2}, \theta$, and may be written in the variational form

$$\delta \int_{t_0}^{t_1} \left(\sum_{j=1}^{2m-2} X_j x_j' + X_{2m-1}\, \theta' \right) dt = 0,$$

to which set of equations must be added the last equation of the first set. But the integrand is positively homogeneous of dimensions unity in $x_1', \cdots, x_{2m-2}', \theta'$, so that θ may be taken as parameter instead of t. Then the variational principle takes the form

$$\delta \int_{t_0}^{t_1} \left[\sum_{j=1}^{2m-2} X_j x_j' + X_{2m-1} \right] d\theta = 0.$$

Hence we obtain a Pfaffian system of even order $2m-2$ only instead of $2m$, in which the coefficients are periodic in a variable θ of period 2π, and the known periodic motion corresponds to the origin in $x_1, \cdots, x_{2m-2}$ space.

By this second method of reduction to a generalized equilibrium problem, the formal difficulties referred to above are avoided.

For these reasons, in dealing with the applications we can restrict attention to the case of generalized equilibrium of stable type as above defined.

2. Stability of Pfaffian systems. Our starting point is furnished by the equations of motion, normalized to terms of an arbitrary degree s by means of an appropriate transformation defined by convergent series, according to the method of the preceding chapter. The equations are thus given the form

$$(1) \qquad \begin{aligned} \frac{dp_i}{dt} &= -\frac{\partial H}{\partial \pi_i}\, p_i + L_{i,s+1}, \\ \frac{dq_i}{dt} &= \frac{\partial H}{\partial \pi_i}\, q_i + M_{i,s+1} \end{aligned} \qquad (i = 1, \cdots m)$$

where we may write

$$H = \sum_{j=1}^{m} \lambda_j p_j q_j + H_4 \cdots + H_{\bar{s}} \qquad (\bar{s} = s \text{ or } s+1),$$

in which H_k involves only the m products $\pi_i = p_i q_i$, of total degree $k/2$ in $\pi_i, \cdots, \pi_m$, while $L_{i,s+1}$, $M_{i,s+1}$ are convergent power series in $p_1, \cdots, q_m$ which commence with terms of degree not lower than $s+1$, the coefficients being of course analytic and periodic in t of period τ.

Suppose that we write

$$u^2 = \sum_{j=1}^{m} p_j q_j.$$

Evidently u can be appropriately regarded as measuring the distance of a point from equilibrium at any instant t; for, in terms of the original real variables $x_1, \cdots, x_{2m}$, the function u^2 is given by a real power series in $x_1, \cdots, x_{2m}$ which begins with a positive definite quadratic form in these variables,

$$u^2 = \sum_{j,k=1}^{2m} a_{jk}(t)\, x_j x_k + \cdots,$$

for all t, whence

$$k \sum_{j=1}^{2m} x_j^2 \leqq u^2 \leqq K \sum_{j=1}^{2m} x_j^2, \qquad K > k > 0,$$

in a certain neighborhood of the origin.

It is obvious then that we can choose N so large that

$$|L_{i,s+1}|,\ |M_{i,s+1}| \leqq N u^{s+1} \quad (i = 1, \cdots, m)$$

within a sufficiently small distance of the origin.

Multiplying the first of the partially normalized equations (1) by q_i, the second by p_i, and adding, we conclude

$$\left| \frac{d\pi_i}{dt} \right| \leqq 2 N u^{s+2} \qquad (i = 1, \cdots, m).$$

From the definition of u, the inequality

$$\left| u \frac{du}{dt} \right| \leqq m N u^{s+2}$$

then follows, so that

$$-mN \leqq \frac{1}{u^{s+1}} \frac{du}{dt} \leqq mN.$$

Integrating from t_0 to t, we deduce from this last inequality

$$\left| \frac{1}{u_0^s} - \frac{1}{u^s} \right| \leqq m s N \,|\, t - t_0 \,|.$$

Now let us ask in how short an interval of time u can exceed $2u_0$. At the corresponding t we obtain

$$\left| \frac{1}{u_0^s} - \frac{1}{2^s u_0^s} \right| < m s N \,|\, t - t_0 \,|,$$

whence obviously since $s \geqq 1$, this cannot happen for

$$|\, t - t_0 \,| \leqq \frac{1}{2 m s N u_0^s}. \tag{2}$$

Hence the minimum time interval which must elapse before the initial distance u_0 can double in magnitude is of the s-th order in the reciprocal distance.

In this same interval of time we obtain

$$\left| \frac{d\pi_i}{dt} \right| \leqq 2^{s+3} N u_0^{s+2},$$

whence by integration

$$|\, \pi_i - \pi_i^0 \,| \leqq 2^{s+3} N u_0^{s+2} \,|\, t - t_0 \,| \quad (i = 1, \cdots, m). \tag{3}$$

Also since H and its partial derivatives are polynomials, we have

$$\left| \frac{\partial H}{\partial \pi_i} - \frac{\partial H^0}{\partial \pi_i} \right| \leqq P \sum_{j=1}^{m} |\, \pi_i - \pi_i^0 \,| \leqq 2^{s+3} m N P u_0^{s+2} \,|\, t - t_0 \,|$$
$$(i = 1, \cdots, m)$$

for π_i, π_i^0 small. On the other hand from the normalized differential equations we find in this interval

$$\left| \frac{dp_i}{dt} + \frac{\partial H}{\partial \pi_i} p_i \right|, \quad \left| \frac{dq_i}{dt} - \frac{\partial H}{\partial \pi_i} q_i \right| \leqq 2^{s+1} N u_0^{s+1}$$
$$(i = 1, \cdots, m).$$

Combining these inequalities with the preceding set, there results

$$\left|\frac{dp_i}{dt}+\frac{\partial H^0}{\partial \pi_i}p_i\right|,\ \left|\frac{dq_i}{dt}-\frac{\partial H^0}{\partial \pi_i}q_i\right|$$
$$\leqq 2^{s+1}Nu_0^{s+1}+2^{s+4}mNPu_0^{s+3}\,|t-t_0|$$

for $i=1,\cdots,m$. These are essentially the same as the following inequalities

$$\left|\frac{d}{dt}(p_i e^{\gamma_i t})\right|,\ \left|\frac{d}{dt}(q_i e^{-\gamma_i t})\right|$$
$$\leqq 2^{s+1}Nu_0^{s+1}+2^{s+4}mNPu_0^{s+3}\,|t-t_0|$$

where $\gamma_i=\partial H^0/\partial\pi_i$ are pure imaginary constants. The fact here made use of, namely that H and its partial derivatives as to π_i are pure imaginaries, can easily be verified: if p_i, q_i, $(i=1,\cdots,m)$, be interchanged and H be changed to its conjugate in (1), these equations are not altered; but this means that the conjugate of H coincides with its negative, i. e. that H is a pure imaginary function. By integration the above inequalities give

$$\begin{aligned}(4)\qquad &|p_i-\overset{0}{p}_i e^{-\gamma_i(t-t_0)}|,\ |q_i-\overset{0}{q}_i e^{\gamma_i(t-t_0)}|\\ &\leqq 2^{s+1}Nu_0^{s+1}\,|t-t_0|+2^{s+3}mNPu_0^{s+3}\,|t-t_0|^2\end{aligned}$$

for $i=1,\cdots,m$.

Now if we return to the convergent power series expressing $x_1,\cdots,x_{2m}$ in terms of $p_1,\cdots,q_m$ and if we replace $p_1,\cdots,q_m$ by

$$\overset{0}{p}_1 e^{-\gamma_1(t-t_0)},\ \cdots,\ \overset{0}{q}_m e^{\gamma_m(t-t_0)}$$

respectively, the series obtained agree with the formal series solutions up to terms of the $(s+1)$-th order in the $2m$ arbitrary constants $p_1^0,\cdots,q_m^0$. But the error committed in so doing is of the order of the differences appearing in (4). Hence if we express $x_1,\cdots,x_{2m}$ by means of the formal series solutions derived from the normal form, broken off after the terms of degree s in the $2m$ arbitrary constants $p_1^0, \cdots, q_m^0$, the error committed will not exceed in numerical value an expression

$$A u_0^{s+1} + B u_0^{s+1} \, | \, t - t_0 \, | + C u_0^{s+3} \, | \, t - t_0 \, |^2$$

during the interval (2), where A, B, C are suitably chosen positive constants.

On account of the fact that s is an arbitrary positive integer in the above inequalities, these can be given a still more simple form. Suppose that $|t - t_0|$ is even more severely restricted than in (2), namely to be of the order at most $(s/3) + 1$ in the reciprocal distance. Then the constitutents of the sum above are clearly of order exceeding $s/3$ in the distance u_0 itself. Consequently if all the terms of the formal series solutions of degree exceeding $s/3$ are discarded, the order of the error will exceed $s/3$. But $s/3$ is arbitrary, whence the conclusion:

If this formal series solution of the generalized Pfaffian equilibrium problem of stable type is written to terms of an arbitrary degree s in the initial values $p_1^0, \cdots, q_m^0$ of the arbitrary constants, the $2m$ trigonometric sums so obtained will have coefficients of at most the first order in u_0, and will represent the coördinates $x_1, \cdots, x_{2m}$ with an error of order u_0^{s+1} at most during a time interval of at least the reciprocal order. Here u_0 represents the distance to the origin for $t = t_0$ in $x_1, \cdots, x_{2m}$ space.

When written out explicitly these trigonometric sums for $x_1, \cdots, x_{2m}$ have the real form

$$A_0 + \sum_j (A_j \cos l_j t + B_j \sin l_j t)$$

where

$$l_i \sqrt{-1} = i_1 \frac{\partial H^0}{\partial \pi_1} + \cdots + i_m \frac{\partial H^0}{\partial \pi_m}, \qquad d = \sum_{j=1}^{m} |i_j| \leqq s,$$

in which $i_1, \cdots, i_m$ are integers, and where A_i, B_i are polynomials in $p_1^0, \cdots, q_m^0$ whose terms are of degree at least d and not more than s.

3. Instability of Pfaffian systems. In the case when some of the multipliers λ_i are real, the situation is entirely altered. If we assume that there are positive and negative multipliers $\pm\lambda_1, \cdots, \pm\lambda_k$, there will be a real k-dimensional, analytic manifold of curves of motion approaching the curve of periodic motion. Points on these curves near to the periodic motion leave its vicinity in a relatively short interval of time. More exactly, the distance will exceed

$$u_0 \, e^{\lambda(t-t_0)},$$

if u_0 denotes the initial distance from the motion at $t = t_0$ and λ is a positive constant less than the least positive multiplier. Similarly if t decreases the distance u_0 may increase in a like manner along a second real analytic manifold of curves.

Evidently such a situation is entirely different from that found in the stable case and is properly termed unstable.

We shall not enter upon a derivation of results of this sort, the first of which were obtained by Poincaré.*

4. Complete stability. The work of section 2 makes it clear that Pfaffian and Hamiltonian systems possess a species of complete formal or trigonometric stability, in case $\lambda_1, \cdots, \lambda_m, 2\pi\sqrt{-1}/\tau$ are pure imaginary quantities without linear relations of commensurability. Let us elaborate this concept of 'complete stability'.

Consider a differential system of even order $2m$,

$$dx_i/dt = X_i(x_1, \cdots, x_{2m}, t) \quad (i = 1, \cdots, 2m), \tag{5}$$

for which the origin is a point of generalized equilibrium. Suppose that for $t = t_0$ the point x_i^0 is at a distance ϵ from the origin. Let T be any fixed time interval, f any positive integer, and $P_s(x_1, \cdots, x_{2m}, t)$ any polynomial with terms of lowest degree s in the coördinates and with coefficients analytic and of period τ in t. If then it is possible always

* For some of the fundamental results see Picard, *Traité d'Analyse*, vol. 3, chap. 1.

to approximate to P_s for $|t - t_0| \leqq T$ with an error less numerically than

$$M\varepsilon^{f+s+1}$$

by a trigonometric sum of order N

$$\sum_{j=0}^{N} (A_j \cos l_j t + B_j \sin l_j t) \qquad (|l_i - l_j| > l > 0),$$

where M, N, l depend only on f and P_s, and where $l_0 = 0$, the equations (5) will be said to be 'completely stable'.

As a very simple example consider the pair of equations

$$dx_1/dt = kx_2, \quad dx_2/dt = -kx_1,$$

of which the general solution is

$$x_1 = A\cos kt + B\sin kt, \quad x_2 = -A\sin kt + B\cos kt,$$

so that the coördinates x_1, x_2 are represented by trigonometric sums of the first order. Any polynomial P_s of degree $s_1 \geqq s$ can also be exactly represented by a sum of order N not exceeding 2^{s_1+1}. Hence the conditions of the definition are satisfied.

The results of section 2 show that in the case of Hamiltonian or Pfaffian systems, there will be complete stability if there is ordinary stability as defined earlier.

This is obvious since the differences $l_i - l_j$ which enter in the trigonometric sums of section 2 are nearly given by a certain limited number of integral linear combinations of the $m+1$ quantities $\lambda_1/\sqrt{-1}, \cdots, \lambda_m/\sqrt{-1}, 2\pi/\tau$, and no such combination vanishes.

In case of complete stability, the solutions of the normalized equations of variation (chapter III, section 5) are limits of trigonometric sums of the specified type, and are trigonometric by the lemma on trigonometric sums of sections 5,6. Hence the multipliers are pure imaginaries.

It is important to establish that this definition of complete stability is independent of the particular coördinates $x_1, \cdots, x_{2m}$ selected. In fact, suppose that the given system is completely stable. Let us make the admissible change of variables

$$\bar{x}_i = \varphi_i(x_1, \cdots, x_{2m}, t) \qquad (i = 1, \cdots, 2m)$$

in which φ_i are analytic in $x_1, \cdots, x_{2m}, t$, vanish at the origin, and are such that the determinant $|\partial \varphi_i / \partial x_j|$ is not 0 there, while the coefficients in φ_i are analytic periodic functions of t of period τ. Then the two variables

$$\bar{\varepsilon} = [\bar{x}_1^2 + \cdots + \bar{x}_{2m}^2]^{1/2}\Big|_{t=t_0}$$

and

$$\varepsilon = [x_1^2 + \cdots + x_{2m}^2]^{1/2}\Big|_{t=t_0}$$

evidently serve equally well to measure the distance from the origin at $t = t_0$, since we have

$$0 < d < \bar{\varepsilon}/\varepsilon < D$$

in the neighborhood of the origin.

Now consider any polynomial $P_s(\bar{x}_1, \cdots, \bar{x}_{2m}, t)$ which can obviously be written

$$P^*(x_1, \cdots, x_{2m}, t) + Q(x_1, \cdots, x_{2m}, t)$$

where P^* is a polynomial in $x_1, \cdots, x_{2m}$ with terms of lowest degree s while Q is given by a power series commencing with terms of degree at least $f + s + 1$. It is clear that the polynomial P^* can be represented by a trigonometric sum of the specified type with an error of order $f + s + 1$ in ε, by the condition for complete stability, while it is clear furthermore that Q is of order $f + s + 1$. Hence it is plain that $P_s(\bar{x}_1, \cdots, \bar{x}_{2m})$ can be represented by the same trigonometric sum in the desired manner. This establishes the complete stability in the new variables.

The mere fact that the multipliers of the system of $2m$ equations of the first order fall into pure imaginary pairs by no means ensures complete stability in the above sense.

A sufficiently simple example is furnished by the pair of equations

$$dx/dt = ky + x(x^2+y^2), \qquad dy/dt = -kx + y(x^2+y^2),$$

in which k is positive, and the fundamental period is taken as 2π. The multipliers are then pure imaginaries, namely $\pm k\sqrt{-1}$. But if the first of these equations be multiplied by $2x$, the second by $2y$, and the two equations so modified be added, there results

$$du/dt = 2u^2 \qquad (u = x^2+y^2),$$

whence, by a further integration

$$u = \frac{u_0}{1-2u_0(t-t_0)}.$$

But, if there were complete stability, it would be possible to find a fixed integer N so large that, for some constant K, the inequality

$$|u - S_N| \leq Ku_0^3$$

held, in which S_N represents a trigonometric expression of order N of the specified type; this follows from the fact that u is a homogeneous polynomial of the second degree in x and y, while u_0 is the squared distance ε^2. This inequality may be written

$$\left|\frac{u-u_0}{u_0^2} - \frac{S_N-u_0}{u_0^2}\right| \leq Ku_0.$$

Now let u_0 approach 0. It is obvious that

$$\lim_{u_0=0} \frac{u-u_0}{u_0^2} = 2(t-t_0).$$

We infer then that

$$\lim_{u_0=0} \frac{S_N-u_0}{u_0^2} = t-t_0.$$

But the expression on the left is a trigonometric sum of order at most N of the specified form, and approaches its limit uniformly. Consequently by the lemma on trigonometric sums considered in sections 5, 6, the limit of this sum is necessarily a sum of the same type. However it is impossible that $2(t-t_0)$ should be so represented. Hence in this case there is not complete stability.

The condition that the multipliers be pure imaginaries has been seen to be necessary for complete stability even if not sufficient. Henceforth we shall assume that, if the m pairs of pure imaginary multipliers be denoted by $\pm\lambda_1, \cdots, \pm\lambda_m$, there are no linear commensurability relations between $\lambda_1, \cdots, \lambda_m$ and $2\pi\sqrt{-1}/\tau$. Of course by so doing certain exceptional cases are excluded which require further study.

For complete stability an infinite number of conditions besides that of pure imaginary multipliers will be found to be requisite.

5. Normal form for completely stable systems. We have already seen that Pfaffian and Hamiltonian systems of equations possess the property of complete stability, in case the characteristic numbers are pure imaginaries. It becomes a very interesting question to determine the most general case in which there is complete stability and to find the characteristics of motion near generalized equilibrium in this case. This we shall do by establishing a suitable normal form for equations of completely stable type.

Since the multipliers $\lambda_1, \cdots, \lambda_m$ are of the stated type, we may transform the variables $x_1, \cdots, x_{2m}$ to $p_1, \cdots, q_m$ by a linear transformation so that the transformed system is

$$dp_i/dt = -\lambda_i p_i + P_i, \quad dq_i/dt = \lambda_i q_i + Q_i \quad (i = 1, \cdots, m)$$

with p_i, q_i conjugate, and P_i, Q_i beginning with terms of at least the second degree.

Now change the variables once more by writing

$$p_i = \bar{p}_i + \bar{\varphi}_{i2}, \quad q_i = \bar{q}_i + \bar{\psi}_{i2} \quad (i = 1, \cdots, m).$$

It is readily found that the equations preserve their form, with the new P_i, Q_i having homogeneous quadratic terms

$$P_{i2} + \sum_{j=1}^{m} \left(p_j \frac{\partial \varphi_{i2}}{\partial p_j} - q_j \frac{\partial \varphi_{i2}}{\partial q_j}\right) \lambda_j - \lambda_i \varphi_{i2} + \frac{\partial \varphi_{i2}}{\partial t},$$

$$Q_{i2} + \sum_{j=1}^{m} \left(p_j \frac{\partial \psi_{i2}}{\partial p_j} - q_j \frac{\partial \psi_{i2}}{\partial q_j}\right) \lambda_j + \lambda_i \psi_{i2} + \frac{\partial \psi_{i2}}{\partial t},$$

respectively. On inspecting these terms and making use of the incommensurability of the multipliers, it appears at once that these new expressions can be made to vanish in one and only one way. In fact let

$$P(t)\, p_1^{\alpha_1} \cdots p_m^{\alpha_m} q_1^{\beta_1} \cdots q_m^{\beta_m} \qquad (\alpha_1 + \cdots + \beta_m = 2$$

be such a term in P_i while the corresponding term in φ_{i2} has a coefficient $\varphi(t)$. By comparison there is obtained the differential equation for φ

$$P(t) + \left[\sum_{j=1}^{m} (\alpha_j - \beta_j)\, \lambda_j - \lambda_i\right] \varphi + \frac{d\varphi}{dt} = 0,$$

which can be satisfied by a periodic function φ of period τ unless the coefficient of φ is an integral multiple of $2\pi \sqrt{-1}/\tau$. This is not possible because of the hypothesis of incommensurability. Moreover the periodic solution is unique (cf. chapter III, section 9).

Thus all of the second degree terms in P_i, Q_i may be removed.

By a precisely similar method all of the third degree terms in P_i, Q_i may be removed by a further transformation

$$p_i = \bar{p}_i + \bar{\varphi}_{i3}, \qquad q_i = \bar{q}_i + \bar{\psi}_{i3} \quad (i = 1, \cdots, m)$$

except when the analogous coefficients

$$\sum_{j=1}^{m} (\alpha_j - \beta_j)\, \lambda_j - \lambda_i, \qquad \sum_{j=1}^{m} (\alpha_j - \beta_j)\, \lambda_j + \lambda_i$$

$$(\alpha_1 + \cdots + \beta_m = 3)$$

vanish. Such exceptional terms will have the form

$$P(t)\, p_i p_j q_j, \qquad Q(t)\, q_i p_j q_j \quad (j = 1, \cdots, m).$$

But even in these terms the functions φ, ψ may be so selected as to make the new coefficients, namely

$$P(t) + \frac{d\varphi}{dt}, \qquad Q(t) + \frac{d\psi}{dt},$$

reduce to constants (cf. chapter III, section 9). Hence we may normalize P_i, Q_i so that

$$\begin{aligned} P_i &= p_i (c_{i1} p_1 q_1 + \cdots + c_{im} p_m q_m) + \cdots, \\ Q_i &= q_i (d_{i1} p_1 q_1 + \cdots + d_{im} p_m q_m) + \cdots, \end{aligned}$$

where the complete terms P_{i3}, Q_{i3} of the third degree appear explicitly written in the right-hand members.

Our next step will be to show that in the event of complete stability we must have the further relations

$$q_i P_{i3} + p_i Q_{i3} = 0 \qquad (i = 1, \cdots, m),$$

i. e., $c_{ij} + d_{ij} = 0$, $(i, j = 1, \cdots, m)$. In order to establish this fact we employ the following lemma which is almost self-evident:

LEMMA ON TRIGONOMETRIC SUMS. If a sequence of trigonometric sums of the type

$$\sum_{j=0}^{N} (A_j \cos l_j t + B_j \sin l_j t) \qquad (|l_i - l_j| > l > 0),$$

with N, l fixed while A_i, B_i, l_i vary except that $l_0 = 0$, approaches a limit $\varphi(t)$ uniformly in some interval, then $\varphi(t)$ is itself a trigonometric sum of order at most N in this interval.

The proof of this simple lemma is deferred to the following section.

Consider the quadratic polynomials $p_i q_i$ in the coördinates. We find from the given equations

$$d(p_i q_i)/dt = q_i P_{i3} + p_i Q_{i3} + \cdots, \quad (i = 1, \cdots, m),$$

where the terms not explicitly indicated are at least of the fifth degree, and where the terms written explicitly are

$$p_i q_i [(c_{i1} + d_{i1}) p_1 q_1 + \cdots + (c_{im} + d_{im}) p_m q_m] \quad (i = 1, \cdots, m).$$

We desire to prove that these terms vanish identically.

Now the differential equations above lead at once to the inequalities

$$|d\pi_i/dt| \leqq K(\pi_1 + \cdots + \pi_m)^2 \qquad (\pi_i = p_i q_i)$$

for $i = 1, 2, \cdots, m$, where $\pi_1, \cdots, \pi_m$ are of course positive or 0. From these last inequalities it follows that we have

$$|du/dt| \leqq mKu^2 \qquad \left(u = \sum_{j=1}^{m} \pi_j\right),$$

and thence

$$|u - u_0| \leqq 4mKu_0^2 |t - t_0| \leqq 2mKTu_0^2$$

for any given interval of time $|t - t_0| \leqq T$, provided that u_0 is sufficiently small. This follows by the methods of section 2. Hence $u - u_0$ is of the second order in u_0 throughout this interval, while the inequalities for $d\pi_i/dt$ show that $\pi_i - \pi_i^0$ is also. Thus

$$q_i P_{i3} + p_i Q_{i3},$$

which is a quadratic polynomial in $\pi_1, \cdots, \pi_m$, differs from its value at $t = t_0$ by terms of the third order in u_0, and the differential equations above give

$$\left| \frac{d\pi_i}{dt} - (q_i^0 P_{i3}^0 + p_i^0 Q_{i3}^0) \right| \leqq L u_0^{5/2} \qquad (i = 1, \cdots, m).$$

By integration there results

$$|\pi_i - \pi_i^0 - (q_i^0 P_{i3}^0 + p_i^0 Q_{i3}^0)(t - t_0)| \leqq LTu_0^{5/2} \quad (i = 1, \cdots, m)$$

in the interval under consideration.

Suppose now that we write

$$p_i^0 = \alpha_i \varepsilon, \quad q_i^0 = \beta_i \varepsilon \quad (i = 1, \cdots, m),$$

$\alpha_1, \cdots, \beta_m$ being m arbitrary pairs of conjugate imaginaries, and suppose that the positive quantity ε approaches 0. The inequality last written in which u_0 is to be regarded as a constant multiple of ε^2, shows that we have

$$\lim_{\varepsilon=0} (\pi_i - \pi_i^0)/\varepsilon^4 = \alpha_i \beta_i [(c_{i1} + d_{i1}) \alpha_1 \beta_1 + \cdots (c_{im} + d_{im}) \alpha_m \beta_m](t - t_0),$$

where the limit is approached uniformly in the interval under consideration. On the other hand π_i can be approximated to by a trigonometric sum of the specified type to terms of order ε^5 in this interval, and consequently $(\pi_i - \pi_i^0)/\varepsilon^4$ can be approximated to terms of order ε. Thus the left-hand member is the uniform limit of a trigonometric sum having the properties specified in the lemma, and must therefore itself be trigonometric. This can only be true if the sums $c_{ij} + d_{ij}$ vanish for all values of i and j, as we desired to prove.

Thus we have to terms of the third order for $i = 1, \cdots, m$,

$$dp_i/dt = -p_i \left[\lambda_i - \sum_{j=1}^{m} c_{ij} p_j q_j \right] + \cdots,$$

$$dq_i/dt = p_i \left[\lambda_i - \sum_{j=1}^{m} c_{ij} p_j q_j \right] + \cdots.$$

Evidently we have begun a process which enables us to remove terms of higher and higher degree in P_i, Q_i except for terms with factors p_i, q_i respectively, and coefficients which are polynomials in the m products $p_i q_i$, one being precisely the negative of the other.

Any completely stable system of equations (5) *may be reduced formally to the normal form*

$$(6) \qquad d\xi_i/dt = -M_i \xi_i, \quad d\eta_i/dt = M_i \eta_i \quad (i = 1, \cdots, m),$$

where $M_1, \cdots, M_m$ are pure imaginary power series in the m variables ξ_i, η_i, i. e.,

$$M_i = \lambda_i - \sum_{j=1}^{m} c_{ij}\,\xi_j\,\eta_j + \cdots \qquad (i = 1, \cdots, m),$$

and ξ_i, η_i are conjugate pairs of variables.

Conversely, if any set of equations have this normal form, the argument of section 2 is available to show that there is complete stability.

6. Proof of the lemma of section 5. Let us consider a sequence of trigonometric sums $\psi(t)$ of the type prescribed in the lemma to be proved. For such a sum we have

$$[D\,(D^2 + l_1^2) \cdots (D^2 + l_N^2)]\,\psi = 0$$

where D indicates ordinary differentiation with respect to t in the symbolic differential operator on the left. Direct integration $2N + 1$ times gives

$$\psi + \left(\sum_{j=1}^{N} l_j^2\right) \int^t \int_0^t \psi(t)\,dt^2 + \cdots$$
$$+ \left(\prod_{j=1}^{N} l_j^2\right) \int_0^t \cdots \int_0^t \psi(t)\,dt^{2N} = P(t)$$

where $P(t)$ is a polynomial of degree at most $2N$.

Now all of the l_i exceed l in absolute value, for by hypothesis

$$|l_i - l_0| = |l_i| \geqq l \qquad (i = 1, \cdots, N).$$

It is clear then that, by suitable choice of a sub-sequence $\psi(t)$, the reciprocals $m_i = 1/l_i$, which are less numerically than $1/l$, will approach limits m_i^* with $|m_i^*| \leqq 1/l$. Any two of the quantities m_i^* will be distinct unless both are 0 of course. Now divide both members of the above integral equation by the product $l_1^2 \cdots l_N^2$, and pass to the limit. Since ψ approaches φ uniformly, we obtain at once

$$\left(\prod_{j=1}^{N} m_j^{*2}\right)\varphi + \cdots + \int_0^t \cdots \int_0^t \varphi(t)\,dt^{2N} = Q(t),$$

where $Q(t)$, being the uniformly approached limit of a sequence of polynomials $P(t)$ of degree not exceeding $2N$, is itself such a polynomial. This leads at once to the conclusion that φ satisfies the linear differential equation with constant coefficients

$$[D(m_1^{*2} D^2 + 1) \cdots (m_N^{*2} D^2 + 1)] \varphi = 0,$$

with general solution a trigonometric sum

$$C_0 + \sum_{j=1}^{N'} [C_j \cos (t/m_j^*) + D_j \sin (t/m_j^*)]$$

where the sum is only extended over those values of j for which m_j^* is not 0. Hence φ is of the stated type.

7. Reversibility and complete stability. It would be possible to show further how intimately the variational principle and the requirement of complete stability are interrelated.† Instead I prefer to follow another direction of thought in order to show that the requirement of complete stability is also very intimately connected with that of reversibility in time of the given differential system, provided that the ordinary definition of reversibility is suitably generalized.

We shall say that a system (5) with generalized equilibrium point at the origin is 'reversible' if when t is changed to $-t$ the system then obtained is equivalent to (5) under the formal group.

By this change of sign of t, the multipliers λ_i are changed to their negatives $-\lambda_i$. Hence it is obvious at the outset that in the reversible case of even order, these multipliers are grouped in pairs, each member of a pair being the negative of the other. We are primarily interested in the case when these multipliers are furthermore pure imaginaries and without linear commensurability relations. For this reason we shall assume that these conditions for first order stability are satisfied.

† See my paper, *Stability and the Equations of Dynamics*, loc. cit.

It is clear that this definition of reversibility is independent of the dependent variables employed. Hence if we have a completely stable system to begin with, we may take it in the normal form (6). The change of t to $-t$ gives a modified system,

$$d\bar{\xi}_i/dt = \bar{M}_i \bar{\xi}_i, \qquad d\bar{\eta}_i/dt = -\bar{M}_i \bar{\eta}_i \qquad (i = 1, \cdots, m)$$

where we introduce the dashes to avoid confusion. But it is possible to pass from one set of equations to the other by the aid of the transformation of the formal group

$$\xi_i = \bar{\eta}_i, \qquad \eta_i = \bar{\xi}_i \quad (i = 1, \cdots, m).$$

Therefore if there is stability of the first order, a necessary condition for complete stability is that (5) *is reversible in the sense of the above definition.*

It remains to prove that this simple necessary condition of reversibility, together with the requirement that the multipliers are of the prescribed type, is also sufficient.

The same process of normalization used in section 5 leads us to the normal form of more general type

$$(7) \quad d\xi_i/dt = U_i \xi_i, \qquad d\eta_i/dt = V_i \eta_i \quad (i = 1, \cdots, m),$$

where U_i, V_i are functions of the m products $\xi_1 \eta_1, \cdots. \xi_m \eta_m$ with initial terms λ_i, $-\lambda_i$ respectively. This may be obtained without the hypothesis of complete stability.

Now if we change t to $-t$ these normalized equations become

$$(8) \quad d\xi_i/dt = -U_i \xi_i, \quad d\eta_i/dt = -V_i \eta_i \quad (i = 1, \cdots, m).$$

These are to be equivalent, by hypothesis, to the original equations (7). It is to be observed that the equations (8) are the same in form as (7) save that the roles of ξ_i and η_i are interchanged, while $-U_i$, $-V_i$ take the place of V_i, U_i respectively.

But it is readily proved that the most general transformation preserving this normal form (7) is of the type

$$\xi_i = \bar{\xi}_i \bar{f}_i, \quad \eta_i = \bar{\eta}_i \bar{g}_i \qquad (i = 1, \cdots, m), \tag{9}$$

where $\bar{f}_i$ and $\bar{g}_i$ are arbitrary power series in the m products $\bar{\xi}_1 \bar{\eta}_1, \cdots, \bar{\xi}_m \bar{\eta}_m$ not lacking constant terms, and with coefficients independent of t.

The fact that these transformations do preserve the normal form is obvious upon direct substitution. In the first place, we note that the inverse relations are of the same type

$$\bar{\xi}_i = \xi_i h_i, \quad \bar{\eta}_i = \eta_i k_i \qquad (i = 1, \cdots, m)$$

with

$$\bar{f}_i h_i = \bar{g}_i k_i = 1 \qquad (i = 1, \cdots, m).$$

Hence we find

$$d\bar{\xi}_i/dt = \bar{U}_i \bar{\xi}_i$$

where

$$\bar{U}_i = \bar{f}_i \left[h_i U_i + \sum_{j=1}^{m} \frac{\partial h_i}{\partial u_j} (U_j + V_j) \xi_j \eta_j \right] \qquad (u_i = \xi_i \eta_i),$$

together with like expressions for $d\bar{\eta}_i/dt$ and $\bar{V}_i$, for $i = 1, \cdots, m$.

In order to establish the fact that this group of transformations is the most general preserving the normal form, we shall proceed step by step.

Consider the terms of the first degree in any such series for $\bar{\xi}_i, \bar{\eta}_i$. These may be written

$$a\xi_i + b\eta_i, \qquad c\xi_i + d\eta_i$$

respectively, so that we must have, for instance,

$$\frac{d}{dt}(a\xi_i + b\eta_i) = a\lambda_i \xi_i - b\lambda_i \eta_i + \xi_i \frac{da}{dt} + \eta_i \frac{db}{dt} + \cdots$$
$$= \lambda_i (a\xi_i + b\eta_i) + \cdots,$$

if the normal form is to be preserved to terms of the first degree only. Hence we infer that b vanishes and a is a constant. Similarly c vanishes and d is a constant, conjugate to a.

Of course this means that the transformations from ξ_i, η_i to $\bar{\xi}_i, \bar{\eta}_i$ are of the specified form as far as the first degree terms are concerned.

Hence the most general transformation which preserves the normal form may be obtained by the composition of the special linear transformation of the group

$$\bar{\xi}_i = a\xi_i, \qquad \bar{\eta}_i = d\eta_i \qquad (i = 1, \cdots, m),$$

and a transformation of the form

$$\bar{\xi}_i = \xi_i + F_i, \qquad \bar{\eta}_i = \eta_i + G_i$$

in which F_i, G_i begins with terms of at least the second degree.

Denote the quadratic terms in F_i and G_i by F_{i2} and G_{i2} respectively. Thus we have to consider

$$\bar{\xi}_i = \xi_i + F_{i2} + \cdots, \qquad \bar{\eta}_i = \eta_i + G_{i2} + \cdots \qquad (i = 1, \cdots, m)$$

with inverse transformation

$$\xi_i = \bar{\xi}_i - \bar{F}_{i2} + \cdots, \qquad \eta_i = \bar{\eta}_i - \bar{G}_{i2} + \cdots \qquad (i = 1, \cdots, m),$$

in which $\bar{F}_{i2}$, $\bar{G}_{i2}$ are merely F_{i2}, G_{i2} respectively with ξ_i, η_i replaced by $\bar{\xi}_i, \bar{\eta}_i$ respectively. We are to determine what is the most general form of F_{i2}, G_{i2} which can preserve the normal form. Now we have

$$\begin{aligned} \frac{d\bar{\xi}_i}{dt} &= U_i\xi_i + \sum_{j=1}^{m} \lambda_j \left(\xi_j \frac{\partial F_{i2}}{\partial \xi_j} - \eta_j \frac{\partial F_{i2}}{\partial \eta_j}\right) + \frac{\partial F_{i2}}{\partial t} + \cdots \\ &\equiv \bar{U}_i \bar{\xi}_i \end{aligned}$$

for $i = 1, \cdots, m$, whence by comparison of second degree terms

$$-\lambda_i F_{i2} + \sum_{j=1}^{m} \lambda_j \left(\xi_j \frac{\partial F_{i2}}{\partial \xi_j} - \eta_j \frac{\partial F_{i2}}{\partial \eta_j}\right) + \frac{\partial F_{i2}}{\partial t} = 0 \quad (i = 1, \cdots, m).$$

The constituent terms which may occur in F_{i2} can be discussed by the methods of section 5, and this leads to the conclusion that F_{i2} must vanish. Likewise G_{i2} is found to vanish. Thus the transformation has the stated form to terms of the second degree inclusive, and it is necessary to consider next a transformation

$$\bar{\xi}_i = \xi_i + F_{i3} + \cdots, \qquad \bar{\eta}_i = \eta_i + G_{i3} + \cdots \qquad (i = 1, \cdots, m)$$

with inverse transformation

$$\xi_i = \bar{\xi}_i - \bar{F}_{i3} + \cdots, \qquad \eta_i = \bar{\eta}_i - \bar{G}_{i3} + \cdots \qquad (i = 1, \cdots, m).$$

Here we are led to m equations

$$-\lambda_i F_{i3} + \sum_{j=1}^{m} \lambda_j \left(\xi_j \frac{\partial F_{i3}}{\partial \xi_j} - \eta_j \frac{\partial F_{i3}}{\partial \eta_j} \right) + \frac{\partial F_{i3}}{\partial t} = \xi_i \Delta U_{i2}$$

$$(i = 1, \cdots, m),$$

where ΔU_{i2} denotes the difference between the second degree components of $\bar{U}_i$ and U_i, when $\xi_1 \eta_1, \cdots, \xi_m \eta_m$ replace $\bar{\xi}_1 \bar{\eta}_1, \cdots, \bar{\xi}_m \bar{\eta}_m$ in $\bar{U}_i$. Thus ΔU_{i2} is a linear function of these m products with constant coefficients. But the method of section 5 shows then that F_{i3} contains a factor ξ_i in every term, these terms being of the form

$$\xi_i \sum_{j=1}^{m} c_{ij} \xi_j \eta_j \qquad (i = 1, \cdots, m).$$

Of course G_{i3} has a corresponding similar form in which η_i appears as a factor.

It follows that the transformation has the stated form to terms of the third degree inclusive. But then the most general transformation can be expressed as one of the specified group followed by a further transformation

$$\bar{\xi}_i = \xi_i + F_{i4} + \cdots, \qquad \bar{\eta}_i = \eta_i + G_{i4} + \cdots \qquad (i = 1, \cdots, m),$$

and we can continue the above method of treatment to the terms of fourth degree and of all higher degrees. Thus we arrive at the conclusion desired that the most general type of formal transformation preserving the normal form is given by (9).

It remains to consider in what cases it is possible to pass from (7) to (8) by a transformation of the type (9), where now we shall introduce dashes so as to distinguish the two sets of variables $\xi_1, \cdots, \eta_m$ and $\bar{\xi}_1, \cdots, \bar{\eta}_m$. If we write $u_i = \xi_i \eta_i$, $\bar{u}_i = \bar{\xi}_i \bar{\eta}_i$, $W_i = U_i + V_i$, we obtain the two associated sets of equations in $u_1, \cdots, u_m$ and $\bar{u}_1, \cdots, \bar{u}_m$

$$(10) \qquad du_i/dt = W_i(u_1, \cdots, u_m)u_i \quad (i = 1, \cdots, m),$$

$$(11) \qquad d\bar{u}_i/dt = -W_i(\bar{u}_1, \cdots, \bar{u}_m)\bar{u}_i \quad (i = 1, \cdots, m),$$

while we have the relations

$$(12) \quad \bar{u}_i = u_i h_i(u_1, \cdots, u_m) k_i(u_1, \cdots, u_m) = u^i l_i(u_1, \cdots, u_m) \quad (i = 1, \cdots, m).$$

Furthermore the constant term in l_i is ϱ_i, a real positive constant, for reasons given above. It is very easy to show that it is impossible that (10) and (11) are related by (12) unless $W_i \equiv 0$.

To begin with, we recall that U_i and V_i have constant terms which are the negatives of one another. In consequence W_i starts off with terms of positive degree r in $u_1, \cdots, u_m$, the aggregate of which we designate by W_{ir}. If we perform the indicated change of variables, we obtain the identities

$$\begin{aligned} \frac{d\bar{u}_i}{dt} &= -\varrho_i W_{ir}(\varrho_1 u_1, \cdots, \varrho_m u_m) u_i + \cdots \\ &\equiv \varrho_i W_{ir}(u_1, \cdots, u_m) u_i + \cdots, \end{aligned}$$

in which only the terms of lowest degree $r+1$ are explicitly indicated. Hence we obtain by comparison

$$W_{ir}(u_1, \cdots, u_m) + W_{ir}(\varrho_1 u_1, \cdots, \varrho_m u_m) = 0.$$

But consider some term of W_{ir}, say

$$c_i u_1^{\alpha_1} \cdots u_m^{\alpha_m} \quad (\alpha_1 + \cdots + \alpha_m = r).$$

This identity yields

$$c_i(1 + \varrho_1^{\alpha_1} \cdots \varrho_m^{\alpha_m}) = 0$$

which is impossible for $c_i \neq 0$. Hence every term of W_{ir} must vanish, which is contrary to assumption. In consequence, we must have $W_i = 0, (i = 1, \cdots, m)$. In other words the requirement of reversibility necessitates that the normal form (8) has the special property characteristic of the case of complete stability.

If there is stability of the first order, reversibility is a sufficient condition for complete stability in the generalized equilibrium problem.

The case of ordinary equilibrium is of course still simpler than that of generalized equilibrium, and the results are entirely analogous.

8. Other types of stability. We have already defined stability of the first order, and complete or trigonometric stability. It was proved in section 2 that, for the equations of dynamics (taken as of Hamiltonian or Pfaffian type), first order stability necessitated complete stability. Other types of stability also possess interest.

In the first place as of the greatest theoretic importance may be mentioned 'permanent stability', for which small displacements from equilibrium or periodic motion remain small for all time. This is the kind of stability of ordinary equilibrium when the potential energy is a minimum. The equations of dynamics are of the type for which this stability *may* obtain, although in general the problem of determining whether or not it does obtain is one of extraordinary difficulty, and constitutes the so-called 'problem of stability'. Thus far the problem has only been solved when a known convergent integral guarantees such actual stability of permanent type.

Another type of stability is that in which these displacements remain small for a very long interval of increasing and decreasing time. A sufficient condition for such 'semi-permanent stability' is the existence of a formal series integral starting off with a homogeneous polynomial of least degree constituting a definite form in the dependent variables. It seems likely that a slight extension of this sufficient condition will turn out to be necessary. Complete stability necessitates semi-permanent stability of course.

Finally a type of 'unilateral stability' in which the displacements remain small for $t > 0$, and in general tend to vanish as t increases indefinitely, has been considered by Liapounoff and others.* It is easy to demonstrate that if the m multipliers possess negative real parts, this kind of stability will obtain. Furthermore it is necessary for this kind of stability that none of these real parts are positive. In the case of the equations of dynamics, however, the real parts of the multipliers can not all be negative, since with every multiplier λ_i is associated its negative. Thus the only possibility of unilateral stability in dynamics is seen to arise when the multipliers are pure imaginaries. In this case the proof of unilateral stability would lead to the proof of permanent stability.

Thus for the problems of dynamics the important types of stability are complete or trigonometric stability, and the permanent stability mentioned above. We shall recur later (chapter VIII) to the important problem of stability concerned with the interrelation of these two types.

* See, for instance, Picard, *Traité d'Analyse*, vol. 3, chap. 8.

CHAPTER V

EXISTENCE OF PERIODIC MOTIONS

1. Role of the periodic motions. The periodic motions, inclusive of equilibrium, form a very important class of motions of dynamical systems. It will be our principal aim in this chapter to consider various general methods by which the existence of periodic motions may be established. In the next chapter a deeper insight into the distribution of the periodic motions is obtained for dynamical systems with two degrees of freedom. Such systems are of the simplest non-integrable type.

The case of a single equation of the first order presents no interest. If the equation is

$$dx/dt = X(x),$$

there will clearly be equilibrium for roots of $X = 0$, while for all other motions there will either be asymptotic approach to one of the equilibrium positions or indefinite increase of $|x|$. Here the equilibrium positions play the central role.

For the case next in order of simplicity there are a pair of equations of the first order. Here the geometric methods of Poincaré* yield the qualitative characteristics of all possible motions, and it is found that the equilibrium positions and periodic motions play the central role. The next following section will be devoted to an example of this type.

When we restrict attention, however, to a pair of equations of Hamiltonian or Pfaffian type (corresponding to a single degree of freedom), there is the special circumstance of an energy integral. Here we assume that the time t does not

* See his paper *Sur les courbes définies par une équation différentielle*, Journal de Mathématiques, ser. 3, vol. 7 (1881), vol. 8 (1882), ser. 4, vol. 1 (1885), vol. 2 (1886).

explicitly appear in the equations; the case when t is involved as a periodic variable is properly to be regarded as of the same degree of generality as that of two degrees of freedom. If we represent the two variables p, q by points in the plane,* the curves of motion will pass one through each point and can only form closed curves or branches open toward infinity in both senses. The corresponding families of periodic motions and unstable motions constitute the totality of motions, except that certain of these curves may contain one or more equilibrium points, in which case there is asymptotic approach to equilibrium in one or both senses.

In the case of dynamical systems of more complicated type it is not clear that the periodic motions take an equally important part. For dynamical systems with only two degrees of freedom, such as are considered in the next chapter, it is, however, almost certain that they continue to play a dominant role. In more complicated cases of still more degrees of freedom, the recurrent motions, introduced in chapter VII, are perhaps to be regarded as the appropriate generalization of the periodic motions, and so likely to become of considerable theoretic importance.

2. An example. The example alluded to in the preceding section is concerned with the rectilinear motion of a particle in a field of force of general type.

More exactly stated, we consider the motion of a particle P of unit mass which moves in a line subject to a force $f(x, v)$ dependent on its space coördinate x and its velocity v.† For the sake of definiteness we shall assume that there is one and only one possible position of equilibrium in the line of motion, and that the motion under consideration is stable in the sense that, for $t > 0$, x and v remain limited in absolute magnitude.

* It is conceivable that p, q are coördinates on a more complicated surface, but we will refer only to the simplest case here.

† For an outline of the treatment here given see my paper, *Stabilità e Periodicità nella Dinamica*, Periodico di Matematiche, ser. 4, vol. 6 (1926).

The usual form for the equation of motion is the single equation

$$d^2x/dt^2 = f(x, dx/dt),$$

where f is a known function which we shall assume to be analytic in the two variables involved. On taking the point of equilibrium at $x = 0$ we have further

$$f(0, 0) = 0; \qquad f(x, 0) \neq 0 \text{ if } x \neq 0.$$

Write $dx/dt = y$ and replace the above equation of the second order by the equivalent pair of equations of the first order

$$dx/dt = y, \qquad dy/dt = f(x, y).$$

These are clearly a pair of equations of the type to which the existence and uniqueness theorems apply.

If we take x, y as the rectangular coördinates of a point Q in the plane, the possible motions of the particle correspond to analytic integral curves, filling the x, y plane, for which

$$dy/dx = f(x, y)/y.$$

The only point curve is the origin O which corresponds to equilibrium. The other curves have everywhere a continuously turning tangent, since not both dx/dt and dy/dt vanish except at the origin. Moreover it is only along the x axis that the slope is infinite.

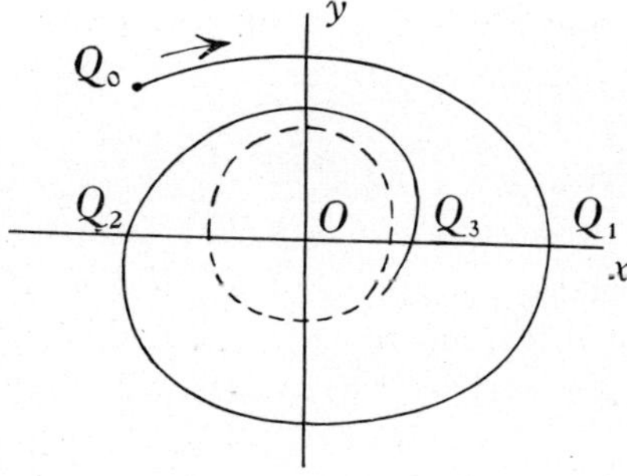

Consider now the particular integral curve corresponding to the given stable motion. Let Q_0 (see figure) be the point of this curve which corresponds to $t = 0$. For definiteness we assume Q_0 to lie in the upper half-plane; the modifications in our statements necessary to make them apply if the point Q_0 lies in the lower half-plane are obvious. Since

$dx/dt = y$ is positive as long as the point Q on the integral curve fails to cross the x axis, Q moves continually towards the right as t increases from 0.

But by hypothesis the point Q lies in a square of sides $2M$ with center at O and sides parallel to the axes.

Hence, while t increases but remains less than any time corresponding to a crossing of the x axis, x, which is also increasing and bounded, must approach a limit $\bar{x}$.

If t approaches a finite limit $\bar{t}$ as x approaches $\bar{x}$, then by the existence theorem the motion can be continued further, but of course not with $y > 0$ according to the definition of $\bar{t}$. Hence in this case $\bar{y}$ must be 0, and the curve crosses the x axis at $(\bar{x}, 0)$. It is to be noted that $\bar{x}$ cannot be zero. Else we should have two solutions, namely the given solution $x(t)$, $y(t)$ and also $x = 0$, $y = 0$, both of which satisfy the initial conditions $x = 0$, $y = 0$ at $t = \bar{t}$, and this would contradict the uniqueness theorem.

If t approaches an infinite limit as x approaches $\bar{x}$, we may show that (x, y) approaches $(0, 0)$. If possible suppose the contrary to be true. It is apparent that, inasmuch as x continually increases but remains less than M in absolute value while y also remains less than M in absolute value, the point (x, y) either approaches a point $(\bar{x}, \bar{y})$, or a strip $(\bar{x}, y)$ where $y_0 \leqq y \leqq y_1$.

But this second possibility would clearly require indefinitely large curvature $\varkappa$ of the integral curve near $(\bar{x}, \bar{y}_0)$ and $(\bar{x}, \bar{y}_1)$, given by

$$\varkappa = \frac{y(f_x y + f_y f) - f^2}{y^2 + f^2}$$

(where f_x, f_y denote partial derivatives), and an indefinitely small value of $y^2 + f^2$. This necessitates of course that $\bar{x}$, $\bar{y}_0$, $\bar{y}_1$ vanish, contrary to hypothesis.

Hence y must approach a value $\bar{y}$ as t approaches infinity.

Now clearly the arc length

$$\int_0^t \sqrt{f^2 + y^2}\, dt$$

will become infinite with t unless (x, y) approaches $(0, 0)$, and indefinitely large arc length would also imply indefinitely large curvature.

It follows then that either Q approaches the origin O from the left as t becomes infinite, or that Q crosses the axis at some point $Q_1 = (x_1, 0)$.

In this second case it is obvious that the point Q, on crossing the x axis at Q_1, starts to move to the left. By the argument employed above, Q will either continue to move to the left and approach O as t becomes infinite, or it must cross the x axis again at $Q_2 = (x_2, 0)$ for the second time. The point Q_2 must, however, lie on the opposite side of O from Q_1 in this case; otherwise $f(\overline{x}_1, 0)$ and $f(\overline{x}_2, 0)$ have the same sign, and the point Q would move downwards at Q_2 as well as at Q_1. It is seen then that Q_1 falls to the right of O, and Q_2 to the left of O.

In fact it follows always that Q_1 is to the right of O, for if Q lies to the left of O, the particle cannot approach O after passing Q_1 since it then moves to the left. Thus a point Q_2 would fall on the same side of O as Q_1.

By repeating this argument indefinitely, we either arrive at a finite number of crossings $Q_1, Q_2, \cdots, Q_n$, alternately to right and left of O, after the last of which Q approaches O, or we find an indefinite set of points $Q_1, Q_2, \cdots$.

From the *analysis situs* of the figure it is apparent that the integral curve must either expand away from O towards a limiting oval about O; or must form such an oval; or must contract towards such an oval (like that indicated in the figure); or contract towards the point O itself. Of course the curve cannot touch or cross itself, by the elementary existence and uniqueness theorems.

Hence we have the following types as the only possible ones for a stable rectilinear motion of a particle in a force field with single equilibrium position:

(a) The particle oscillates indefinitely often back and forth past the equilibrium position with expanding amplitude, and approaches a periodic motion asymptotically.

(b) The particle oscillates periodically about the equilibrium position.

(c) The particle oscillates indefinitely often back and forth with decreasing amplitude and approaches a periodic motion.

(d) The particle oscillates back and forth a finite or an infinite number of times and approaches the equilibrium position.

(e) The particle is in the equilibrium position.

A very clear idea of all the possible motions of the particle under an arbitrary law of force of this type is readily obtained, on the basis of the above discussion.

Let us consider the ordered set of distinct closed curves in the x, y plane corresponding to the periodic motions. All of these must evidently enclose the origin which can properly be regarded as the innermost curve of the set.

Any other curve of motion may lie between a pair of the periodic curves in which case it will continue between them always and so be stable. The particle then approaches one of these enclosing periodic motions asymptotically as t increases indefinitely and the other as t decreases indefinitely.

In the only alternative case the curve will lie outside of the outermost curve of the set of periodic motions, and will clearly be stable in one and only one sense of time t and will approach the corresponding periodic motion asymptotically in this sense.

3. The minimum method. Suppose now that we have a Lagrangian dynamical problem,

$$\delta I = \delta \int_{t_0}^{t_1} L\,dt = 0,$$

where L is a function of the space coördinates $q_1, \cdots, q_m$ and of the velocities $q_1', \cdots, q_m'$, being quadratic in these last variables. There is an integral of energy, namely

$$L_2 - L_0 = \text{const.}$$

By adding an appropriate constant to L_0, we may assume that the arbitrary constant vanishes for the given motion.

We propose to simplify the given problem by making use of the integral of energy which may then be written

$$L_2 - L_0 = 0.$$

As has already been seen, it is possible to give an alternative variational formulation of the problem

$$\delta \int_{t_0}^{t_1} 2 \sqrt{L_0 L_2} + L_1)\, dt = 0$$

(see chapter II, section 3), in which the integrand is positively homogeneous of dimensions unity, so that the value of the integral is independent of the parameter t employed along the path of integration.

Now the underlying variables here are $q_1, \cdots, q_m$. But it is not important which particular set of variables is employed, since an arbitrary one-to-one, analytic transformation does not affect the validity of the variational principle. It is important, however, to require that the sets of coördinate values $q_1, \cdots, q_m$ form a certain analytic manifold M of known connectivity. Accordingly we assume that the coefficients in L are analytic in $q_1, \cdots, q_m$ over this manifold when $q_1, \cdots, q_m$ are properly chosen. Let us furthermore assume that $4 L_0 L_2 - L_1^2$, which is a homogeneous quadratic form in the velocities, is a positive definite form. The expression $ds^2 = L_0 L_2 \, dt^2$ may be considered as the element of arc on the 'characteristic surface' M.

Let l denote any closed curve in M which cannot be continuously deformed to a point. Here M is taken to be multiply connected in the sense of linear connectivity.

Suppose further that under such deformation the integral I taken around the curve l increases indefinitely if l does not remain wholly on the finite part of the manifold, and also that I exceeds a certain positive constant I_0 for any choice of l. There will then be a positive lower bound for I along these curves.

It is intuitively apparent that this bound will be attained, and will yield a closed curve corresponding to a periodic motion. We shall not attempt here to enter any of the logical details but merely to state the result.*

Given a Lagrangian dynamical problem of this type with

$$L = L_0 + L_1 + L_2,$$

and such a circuit l not deformable to a point on the characteristic surface, then, for the prescribed value c of the energy constant, i. e., of

$$L_2 - L_0 = c,$$

there exists a periodic motion of the same type as l for which

$$I = \int_l (2\sqrt{(L_0 + c)\, L_2} + L_1)\, dt$$

is an absolute minimum.

If $L_1 = 0$, so that the problem is reversible, the integral I becomes the arc length s on the characteristic surface, and the periodic motion corresponds to a closed geodesic of the given type.

In the case of two degrees of freedom ($m = 2$), this minimum method yields only those periodic motions of unstable type for which the two non-vanishing multipliers are real.† Doubtless analogous results hold for any number of degrees of freedom, and can be obtained by means of classical methods in the calculus of variations.

If the energy constant c varies, it is almost obvious that these periodic motions of minimum type vary analytically. Here is an instance then of the analytic continuation of a periodic motion with variation of a parameter (see section 9).

4. Application to symmetric case. There is one case in which the direct application of the minimum method fails, namely that in which there is no circuit l on the characteristic

* Cf. my paper, *Dynamical Systems with Two Degrees of Freedom*, Trans. Amer. Math. Soc., vol. 18 (1917), where the minimum method is developed more fully, and the important antecedent papers by Hadamard, Whittaker, Hilbert and Signorini are referred to.

† Cf. my paper (loc. cit), section 14.

surface not reducible to a point. It is interesting to observe that even here a slight modification of the minimum method may be applicable. This will be possible in those cases when the dynamical problem is 'symmetric', in the sense that the points of the characteristic surface may be grouped in distinct symmetric pairs, such that I has the same value along any curve and along its symmetric image. When this property is satisfied and the coördinates $q_1, \cdots, q_m$ of each point of a pair are taken to be locally the same, then L_0, L_1, L_2 are also the same at symmetric points.

To illustrate the idea involved, let the surface M be thought of as lying in ordinary space, and symmetric in the origin (but not passing through it), so that if x, y, z are the coördinates of a point of M then $-x, -y, -z$ are the coördinates of the symmetric point of M. Of course M is taken to be connected and as having the properties previously specified; in particular, M may be a convex surface symmetrical in the origin. The integral I may be thought of as ordinary arc length along a curve in the surface M.

Now suppose any closed curve $ABCDA$ is drawn in M such that ABC is the image of CDA, with A and C as symmetrical points. Let this curve l be continuously deformed in any manner, but with the condition that it is always to be composed of two symmetrical arcs ABC, CDA.

The integral I along l will then possess an absolute minimum which will be attained along some curve of this type. In fact we need only regard symmetric points as identical and consider I along the closed curve ABC on the modified manifold M' so defined.

If a Lagrangian problem of this type admits of symmetry in the sense specified, and if l is any symmetric circuit on the characteristic surface M, there will exist a symmetric periodic motion of the same type as l, for which I over any l is an absolute minimum.

In particular let there be a closed m-dimensional analytic surface with the connectivity of a hypersphere, which lies in $(m+1)$-dimensional space and is symmetric in the origin.

The result above is at once applicable and indicates the existence of at least one closed geodesic without multiple points.

More generally, if the Lagrangian problem of this type admits of an analytic transformation T into itself, whose k-th iterate is the identical transformation, and if l is a circuit invariant under T and not deformable to a point on M, there will exist a periodic motion of the same type as l.

5. Whittaker's criterion and analogous results. Hitherto we have dealt with Lagrangian dynamical problems possessing characteristic surfaces M without any boundary save at infinity. It is, however, very simple to extend our results to the case when M has one or more analytic $(m-1)$-dimensional boundaries, provided that the unique short geodesic arc joining any ordered pair of nearly points within M also lies within M. A boundary will be termed 'convex' in case it has this property.

In fact the minimizing curve in M is then either a closed extremal, in which case it is obviously not tangent to any of these boundaries and so lies within M, or it is composed of a finite or infinite number of extremal arcs whose vertices must lie on these boundaries of course. But, by the very definition of the convex boundary, boundary vertices cannot occur on the minimizing curve. In fact a short arc AVB of such a minimizing curve with vertex at V can be replaced by the shorter single extremal arc AB, which must lie wholly within M. This is absurd.

The surface M, defined in the preceding section, may be assumed to possess any number of finite convex boundaries in addition to the possible boundaries at infinity, without affecting the existence theorems there obtained.

The original criterion of Whittaker referred to the reversible case of two degrees of freedom when M was a ring. Here the result obtained is that there is a periodci motion of minimum type which makes a single circuit of the ring.*

* Cf. my paper, loc. cit., sections 10–13.

6. The minimax method. By means of the 'minimax' method it is possible to establish the existence of still further periodic motions.

Perhaps the simplest illustration of the method is obtained through the consideration of the geodesics on a torus-shaped surface in ordinary three-dimensional space. Evidently the minimum method explained above yields at least one closed geodesic having the type of any closed curve on the torus not deformable to a point. Now let a closed curve l be moved in any way from this minimizing geodesic back to the same position, while at least one of the two angular coördinates increases by $2k\pi$. It will be necessary certainly to increase the length of l during this movement of l, and there will be a least upper bound of length l^*, necessary in order that the movement be possible. At some intermediate stage then the varying curve will actually be of this length and will be taut. This position of l corresponds to a closed geodesic. Clearly it is not possible to deform all nearby curves of length less than l^* into one another or l^* would not be the minimax specified. This property is characteristic of geodesics of minimax type.

The above treatment is intuitive. However a rigorous treatment can be given.†

More generally we are led to the following conclusion: *A Lagrangian problem restricted as in section* 4 *and with* $k > 1$ *periodic motions of minimum type associated with a closed curve* l, *will necessarily possess at least* $k-1$ *further periodic motions of minimax type associated with this same circuit.*

If we omit consideration of all exceptional cases and use an intuitive form of reasoning, this more general principle may be made plausible as follows. Let $I_1, \cdots, I_k$ be the values of the integral I along the k periodic motions of minimum type which are known to exist, and let I^* be so large that it is possible to deform a curve l from any one of

† Cf. my paper, loc. cit., sections 15–19, for a development of the minimax method in the case of two degrees of freedom.

the corresponding curves $l_1, \cdots, l_k$ into any other without I becoming as large as I^*. For definiteness let us suppose that $I_1, \cdots, I_k$ are in increasing order of magnitude.

Let u be a variable parameter and consider the closed curves l of the given type for which $I<u$. For $u<I_1$ (the absolute minimum) there are no curves but, as u increases through I_1, closed curves not differing much from the curve for I_1 are admitted, and the more u increases the more extensive the variation of l about the curve for I_1 may be with $I<u$. Similarly as u increases through I_2 a new isolated set of curves l in the neighborhood of the curve for I_2 come into existence, for which $I<u$. And as u increases finally through I_k a last k-th set of curves comes into existence about the curve for I_k.

Now as u increases some two of these k sets of curves may unite, i. e. it becomes possible to deform a curve l from the curve for one I_α into a curve for I_β. There will then be a least value of u for which this is possible, and a corresponding periodic motion of minimax type. Each time that such a union takes place the number of sets of curves l with $I<u$ is diminishes by 1.

But when $u=I^*$ there is only one common group, so that $k-1$ unions have taken place. This is what we desired to prove.

It is not difficult to show that, unless the periodic motion of minimax type is multiple, only two sets of curves can coalesce at one of these critical values of u.

When the characteristic surface admits of discrete transformations into itself, an exceptional case arises in which the periodic motions of minimum type are to be counted more than once. This is the case of the torus alluded to above.

It is also of interest to observe that when a curve l is regarded as described k times, $k>1$, the motions of minimum type remain so, but the motions of minimax type associated with them will not be the same as in the case $k=1$, but will be distinct from these.

The general situation here requires further study.

7. Application to exceptional case. The case of the m-dimensional Lagrangian problem when the characteristic surface can be set into one-to-one analytic correspondence with the hypersphere is of exceptional interest, but the minimax method outlined above is not directly applicable since there are no closed curves l, not deformable to a point, from which to start. Nevertheless the existence of a periodic motion of minimax type may be established.

In order to make the reasoning as concrete as possible we shall direct attention to the reversible geodesic problem, although it is clear that the reasoning applies equally well for a Lagrangian problem of the kind treated in the preceding sections, with characteristic surface homeomorphic with the hypersphere.

Our first step will be to define what is meant by a 'covering' of the surface. In the case of a two-dimensional surface let the surface of the sphere be set into one-to-one analytic correspondence with the given surface. The small circles on the sphere in planes perpendicular to some axis are evidently carried into a set of closed analytic curves covering the given surface, two of these being point-curves. Thus we may conceive of the spherical surface as being distorted analytically to form a covering of the given surface M by means of this set of closed curves. The points of the covering can then be specified by two angular coördinate functions θ, φ on the surface where θ and φ represent colatitude with respect to the given axis and longitude respectively. The given closed curves correspond to $\theta =$ const., while φ varies from 0 to 2π. The coördinate θ ranges from 0 to π only, with the two extreme values corresponding to the point curves.

Now conceive of this covering as continuously deformed. This means that each point of the covering is carried by continuous variation into nearby points, while the curves of the covering go over into new curves. It is obvious that such a covering will always actually cover each point at least once and cannot reduce to a point.*

* The suggestion for a proof may be found in a footnote, p. 246 of my article, loc. cit., and this proof extends readily to the m-dimensional case.

Similarly for the m-dimensional case, we introduce a system of small circles on the hypersphere

$$x_1^2 + \cdots + x_{m+1}^2 = 1$$

($x_1, \cdots, x_{m+1}$ rectangular coördinates) with equations

$$x_3 = x_3^0, \cdots, x_{m+1} = x_{m+1}^0, \quad x_1^2 + x_2^2 = 1 - x_3^{0^2} \cdots - x_{m+1}^{0^2}.$$

Here the null circles of the set are in one-to-one, continuous correspondence with an $(m-1)$-dimensional hypersphere.

The image of this system of circles leads to an analytic covering of the given characteristic surface M. The points of the covering can then be specified by suitable coördinates, and we may conceive of the covering as continuously varied. It is obvious that such a covering will always cover each point of M at least once.

Now there is a maximum length L^* for any image of a circle, and there can be selected a distance d such that two points at geodesic distance not greater than d from each other in M are connected by a unique minimizing geodesic of length $\delta < d$. Let n be the positive integer such that

$$\frac{L^*}{n} < d \leqq \frac{L^*}{n-1}.$$

On the image of any circle select the point P_1 as the point which corresponds to $\varphi = 0$ and divide the curve into n arcs

$$P_1 P_2, P_2 P_3, \cdots, P_n P_1,$$

of equal length $< d$. Let a point Q_i move from P_i to P_{i+1} ($P_{n+1} = P_1$) on each such arc in such wise that each arc is constantly divided proportionately. Consider the short geodesic arc $P_i Q_i$ and the arc $Q_i P_{i+1}$ of $P_i P_{i+1}$ during this process. At the outset the combined arc is $P_i P_{i+1}$ while at the conclusion it has been varied continuously to the geodesic arc $P_i P_{i+1}$.

In this way we see that it is possible by continuous variation to replace the covering with the given images of the circles by a covering of closed curves made up of geodesic arcs $P_1 \cdots P_n P_1$. Furthermore it is clear that the maximum length of any curve of the new covering cannot exceed L^*, while the maximum length of any component geodesic arc is less than d.

This constitutes our first step in a sequence of continuous variations of the given covering. The second step is to divide each curve of geodesic arcs in n equal parts starting with the midpoint of $P_1 P_2$. Thus points $Q_1, \cdots, Q_n$ are obtained, and a second modified covering made up of geodesic arcs $Q_1 Q_2, \cdots, Q_n Q_1$, each of length less than d, while the closed curves are of length less than L^*. The modification can be affected by essentially the same device as before.

The process of successive n-section and variation thus commenced can be indefinitely continued. At each stage the individual arcs are of length less than d, and the total length of each curve is less than L^*. Furthermore the effect of each step is to diminish (or at least not to increase) the length of the curve.

It is conceivable that some of the intermediate curves reduce to points during the process, but this would in no way invalidate the reasoning here made. However it is not possible that the maximum length of every curve becomes less than d. This may be seen readily. In the contrary case the curves $P_1 \cdots P_n P_1$ of geodesic arcs of the corresponding covering can all be reduced to point-curves as follows: let each point Q move toward P_1 along the unique short geodesic joining it to P_1, in such wise that proportional parts of this distance are simultaneously described by every point P. In this way the m-dimensional covering becomes $(m-1)$-dimensional at most, and so cannot pass through all points of M, which is absurd.

In connection with this last step it should be noted that the process adopted leaves the set of points P_i an analytic $(m-1)$-dimensional manifold at each stage, so that no difficult point-set questions connected with dimensionality arise.

We infer that the maximum length L_p at the p-th stage diminishes as p increases, and approaches a positive lower limit $L > d$.

It is now easy to prove that any corresponding sequence of geodesic arcs

$$P_1^{(p)} \cdots P_n^{(p)} P_1^{(p)} \qquad (p = 1, 2, \cdots)$$

for which the length is this maximum L_p will contain a closed limiting geodesic of length L. In fact we shall be able to establish this fact at once on the basis of the following lemma:

LEMMA. If any closed curve of n equal arcs $P_1 P_2, \cdots, P_n P_1$ each of length $\leqq d$, and of total length $\geqq d$, is modified to the curve of geodesic arcs $P_1 \cdots P_n P_1$, and then to the series of geodesic arcs $Q_1 \cdots Q_n Q_1$ joining the midpoint Q_1 of $P_1 P_2$ with the points $Q_2, \cdots, Q_n$ of n-section, and if the exterior angle between the geodesics $P_i P_{i+1}$ at some vertex exceeds $\delta > 0$, then the difference between the length of the initial and final curve exceeds a specifiable positive constant dependent only on δ.

In the first place we note that the two steps each decrease total length. Hence the proposition will certainly be true unless the first step changes the total length only very slightly. Since n is fixed one for all, this means that each geodesic arc $P_i P_{i+1}$ is substantially the same in length as the original arc $P_i P_{i+1}$. By Osgood's theorem in the calculus of variations, the original arc must be very near to the modified geodesic arc, and these latter arcs are nearly equal also. Hence the points of Q_i of n-section of the geodesic arcs fall very near to the points half way between the points P_i of n-section on the original curve. In consequence it appears that if the exterior angle at any P_{i+1} exceeds a specified δ the sum of the geodesic arcs $Q_i P_{i+1}$ and $P_{i+1} Q_{i+1}$ will exceed the geodesic arc $Q_i Q_{i+1}$ by a specifiable positive quantity. This would lead to the desired conclusion.

Thus an outline of the proof has been given. Obviously the proof is of such a character that a full statement of all the questions of uniformity involved would be lengthy, although the outline sufficiently indicates the general direction of procedure.

On the basis of the lemma our discussion is at once completed. In fact the lengths of the sucessive curves specified would decrease indefinitely often by a specifiable amount unless the exterior angles at the vertices of the geodesic arcs approach 0 uniformly. Hence these exterior angles approach 0. But the points P_1^p have at least one limit point P, and the directions of the geodesics have a limiting direction, so that there is a limiting geodesic which is clearly closed and of length L precisely.

If the m-dimensional characteristic surface M is homeomorphic with the m-dimensional hypersphere, there exists at least one periodic motion obtained by the process specified above.

It is natural to expect that such a motion is of minimax type, but we shall not attempt to establish this conjecture here. In the simplest case of two degrees of freedom this conjecture holds true.

8. The extensions by Morse. The methods of minimum and minimax suffice only to give certain of the periodic motions. Remarkable recent work by Morse* indicates with a high degree of probability that all of the types of periodic motions can be discovered by suitable extension of these methods, based on a deeper use of the principles of *analysis situs.* Moreover the number of periodic motions of the various types (the minimum and minimax types being the simplest of a series) are characterized by certain interrelations discovered by Morse, although so far only explicitly developed by him as to apply to dynamical systems with two degrees of freedom in the neighborhood of a periodic motion.

9. The method of analytic continuation. The method of analytic continuation of Hill and Poincaré starts with a

* See his paper *Relations Between the Critical Points of a Function of n Independent Variables,* Trans. Amer. Math. Soc., vol. 27 (1925), and a forthcoming paper, *A Theory of the Ordinary Problem of the Calculus of Variations in the Large.*

known periodic motion and then obtains an analytic continuation of it with variation of a parameter c.

For the sake of definiteness we consider a system of Hamiltonian equations

$$\frac{dp_i}{dt} = -\frac{\partial H}{\partial q_i}, \qquad \frac{dq_i}{dt} = \frac{\partial H}{\partial p_i} \quad (i = 1, \cdots, m),$$

in which H is analytic in $p_1, \cdots, q_m, t, c$, and periodic in t of period 2π. Furthermore we suppose that the origin is a point of generalized equilibrium for $c = 0$.

By proper preliminary linear change of variables of the type made in chapter III, section 6, we may take H in the normal form

$$H = -\sum_{j=1}^{m} \lambda_j p_j q_j + H_3 + \cdots$$

for $c = 0$, in case $\lambda_1, \cdots, \lambda_m$ are distinct.

Now the general solution of the given system may be written

$$p_i = p_i(p_1^0, \cdots, q_m^0, t, c), \qquad q_i = q_i(p_1^0, \cdots, q_m^0, t, c)$$

for $i = 1, \cdots, m$, where $p_1^0, \cdots, q_m^0$ denote the values of $p_1, \cdots, q_m$ respectively at $t = 0$.

The condition for periodicity is then contained in the system of $2m$ simultaneous equations

$$p_i(p_1^0, \cdots, q_m^0, 2\pi, c) = p_i^0, \qquad q_i(p_1^0, \cdots, q_m^0, 2\pi, c) = q_i^0$$
$$(i = 1, \cdots, m).$$

Here all the variables are involved analytically as was established in chapter I. But this system of equations admits of the solution

$$p_1^0 = \cdots = q_m^0 = 0$$

for $c = 0$, by hypothesis. Hence there will be a unique solution $p_1^0, \cdots, q_m^0$, analytic in c, provided that the functional determinant

$$\begin{vmatrix} \frac{\partial p_1}{\partial p_1^0} - 1, & \frac{\partial p_1}{\partial p_2^0}, & \cdots, & \frac{\partial p_1}{\partial q_m^0} \\ \cdot & \cdot & \cdot & \cdot \\ \cdot & \cdot & \cdot & \cdot \\ \frac{\partial q_m}{\partial p_1^0}, & \frac{\partial q_m}{\partial p_2^0}, & \cdots, & \frac{\partial q_m}{\partial q_m^0} - 1 \end{vmatrix}$$

does not vanish for $t = 2\pi$, $c = 0$. But the $2m$ functions

$$\frac{\partial p_1}{\partial p_i^0}, \cdots, \frac{\partial q_m}{\partial p_i^0}$$

$(i = 1, \cdots, m)$ form m solutions of the equations of variation, as do also

$$\frac{\partial p_1}{\partial q_i^0}, \cdots, \frac{\partial q_m}{\partial q_i^0}$$

$(i = 1, \cdots, m)$. In addition these reduce to 0 at $t = 0$, except for $\partial p_i / \partial p_i^0$ and $\partial q_i / \partial q_i^0$ which are 1 for $i = 1, \cdots, m$.

The known nature of the terms in H for $c = 0$ yields as equations of variation

$$d y_i / d t = \lambda_i y_i, \qquad d z_i / d t = -\lambda_i z_i \qquad (i = 1, \cdots, m),$$

where y_i, z_i correspond to p_i, q_i. Hence the solutions above are of the following explicit forms

$$\begin{matrix} e^{\lambda_1 t}, & 0, & \cdots, & 0, \\ 0, & e^{\lambda_2 t}, & \cdots, & 0, \\ \cdot & \cdot & \cdot & \cdot \\ 0, & 0, & \cdots, & e^{-\lambda_m t}, \end{matrix}$$

where the only non-zero elements are the diagonal terms, namely

$$e^{\lambda_1 t}, \cdots, e^{\lambda_m t}, e^{-\lambda_1 t}, \cdots, e^{-\lambda_m t}.$$

Consequently the determinant written above reduces to

$$\prod_{j=1}^{m} (e^{2\pi\lambda_j} - 1)(e^{-2\pi\lambda_j} - 1),$$

and cannot vanish unless some λ_i is an integral multiple of $\sqrt{-1}$.

Under these circumstances then we have an analytic family of solutions

$$p_1(t, c), \cdots, q_m(t, c),$$

periodic of period 2π in t, as we desired to prove.

These restrictions may be notably lightened. In the first place, when not all the multipliers $\lambda_1, \cdots, \lambda_m$ are distinct, a similar normal form of solution exists. For example, take $\lambda_1 = \lambda_2$, while taking $\lambda_2, \cdots, \lambda_m$ distinct from each other and λ_1. In general the first and second solutions now have the forms

$$0, e^{\lambda_1 t}, 0, \cdots, 0,$$
$$e^{\lambda_1 t}, te^{\lambda_1 t}, 0, \cdots, 0,$$

and the $(m+1)$-th and $(m+2)$-th solutions likewise

$$0, e^{-\lambda_1 t}, 0, \cdots, 0,$$
$$e^{-\lambda_1 t}, te^{-\lambda_1 t}, 0, \cdots, 0$$

respectively. If the first and second rows are interchanged as well as the $(m+1)$-th and $(m+2)$-th rows, it appears that the final determinant will have vanishing elements on the lower side of the diagonal and will not itself vanish unless some λ_i is an integral multiple of $\sqrt{-1}$.

This is an entirely general result, namely that analytic continuation is possible so long as there is not a multiplier λ_i which is an integral multiple of $\sqrt{-1}$.

But such a multiplier indicates neither more nor less than the presence of a solution of the equations of variation with the period 2π of the given motion.

Let us define the generalized equilibrium point as 'simple' when there is no solution of the equations of variation with the period of the generalized equilibrium point, and otherwise as 'multiple'.

Analytic continuation of the generalized equilibrium is possible so long the equilibrium is simple.

By a change of variables

$$p_i = p_i(t, c) + P_i, \qquad q_i = q_i(t, c) + Q_i \quad (i = 1, \cdots, m),$$

we obtain modified Hamiltonian equations

$$\frac{dP_i}{dt} = -\frac{\partial H^*}{\partial q_i}, \qquad \frac{dQ_i}{dt} = \frac{\partial H^*}{\partial p_i} \quad (i = 1, \cdots, m)$$

where

$$H^* = H + \sum_{j=1}^{m} (p'_j Q_j - q'_j P_j).$$

These are of the same type as before, but have a generalized equilibrium point at the origin *for all small values of* c. We have written p'_i, q'_i for the time derivatives of $p_i(t, c)$, $q_i(t, c)$ respectively.

The formal series solutions of a system of this type will of course also involve the parameter c. It is formal series of this type which are often useful in the applications; and the vanishing of the parameters such as c may correspond to a special integrable case of the dynamical problem when the periodic motion from which we start admits of explicit determination.

10. The transformation method of Poincaré. Sometimes a dynamical problem can be given a striking change of form. In fact the solutions of n differential equations of the first order, with right-hand members independent of the time t, can be represented by the steady motion of an n-dimensional fluid, of which a moving point has the dependent variables as coördinates. Now suppose that a closed, $(n-1)$-dimensional, analytic surface S can be constructed in this 'manifold of states of motion', such that every stream line cuts S at least once within any fixed interval of time τ, sufficiently large, and always in the same sense. Then S may be called a 'surface of section'. If a point P of S be followed along the stream line through P in the sense of increasing time, it intersects S again at a first subsequent point P_1. Thus there is defined a one-to-one, direct, analytic transformation of the surface of section into itself, namely the transformation T which takes each point P into the point P_1.

In this manner it may be possible to associate the given dynamical problem with a discrete transformation T of a closed $(n-1)$-dimensional surface into itself. Properties of the motions are then mirrored in properties of T. For example, the periodicity of a motion represented by a closed curve in the manifold of states of motion meeting S in $P, P_1, \cdots, P_{k-1}$ is reflected in the symbolic equations

$$P_1 = T(P),\ P_2 = T(P_1), \cdots, P = T(P_{k-1}),$$

so that $P, P_1, \cdots, P_k$ are all invariant points of S under the k-th iterate of T. Conversely, if P is so invariant, there is a corresponding periodic motion, represented by a closed curve meeting S in the k points $P, T(P), \cdots, T^{k-1}(P)$.

A surface of section in this sense will only exist if there is an angular variable φ in the manifold of states of motion which may be so defined as to constantly increase along each stream line. The necessity of this condition may be seen as follows. If a surface of section S exists, let φ be defined as 0 on this surface and as $2\pi t/\tau$ at any other point P, where τ is the complete time interval necessary to pass from S to S along the stream line on which P lies. Evidently φ is an analytic function of position which increases along every stream line by exactly 2π between successive intersections with S. Hence an angular variable φ exists. Conversely if such a variable φ exists, then $\varphi = 0$ will yield a surface of section.

A necessary and sufficient condition for a closed surface of section is the existence of an angular variable φ in the manifold of states of motion which constantly increases along every stream line.

More explicitly stated, there must exist a differential inequality

$$\frac{d\varphi}{dt} = \sum_{j=1}^{n} \Phi_j X_j > 0$$

in which Φ_i satisfy integrability conditions,

$$\frac{\partial \Phi_i}{\partial x_j} = \frac{\partial \Phi_j}{\partial x_i} \qquad (i, j = 1, \cdots, n),$$

and $X_1, \cdots, X_n$ denote the right-hand members of the differential equations as usual.

An extremely interesting type of surface of section possessing boundaries can be found in certain dynamical problems. Here the boundaries of S are closed analytic $(n-2)$-dimensional manifolds of stream lines, and every stream line not on these boundaries of S cuts the interior of S at least once within any interval of time τ, sufficiently large, and always in the same sense.

In the case $n=3$, the surface of section is two-dimensional, and its boundaries may then be the closed curves corresponding to a single periodic motion. Now in the case of a Hamiltonian or Pfaffian dynamical problem with two degrees of freedom, the use of the energy integral reduces the order to $n=3$. For such problems there seem in general to exist surfaces of section, as will appear in the next chapter.* The example of the next section illustrates the possibility of such a surface of section when there are more than two degrees of freedom, In such a case, however, it is necessary to make use of an $(n-2)$-dimensional, analytic, closed surface made up of stream lines, such as do not appear in general to exist.

Another case of very decided interest is that of an open analytic surface of section such as Koopman† has obtained in the exterior case of the restricted problem of three bodies.

It is evident in all of these cases that by the reduction to a transformation problem, the determination of the periodic motion is made to hinge upon that of the invariant points of a surface of section S under a transformation T and its iterates.

* Cf. my paper, loc. cit., sections 22–29.

† *On Rejection to Infinity and Exterior Motion in the Restricted Problem of Three Bodies,* Trans. Amer. Math. Soc., vol. 29 (1927).

11. An example. An example showing that such surfaces of section may exist for Hamiltonian problems of more than two degrees of freedom is afforded by the following dynamical problem:

A particle P in a conservative field of force in space moves in such wise that the force has always a positive component towards a fixed plane for points outside of that plane.

Here the equations of motion form a system of the sixth order and may be written

$$dx/dt = x', \qquad dy/dt = y', \qquad dz/dt = z',$$
$$dx'/dt = -\partial U/\partial x, \quad dy'/dt = -\partial U/\partial y, \quad dz'/dt = -\partial U/\partial z,$$

where x, y, z are the rectangular coördinates of P in space, and where $z = 0$ may be taken as the fixed plane. Also $\partial U/\partial z$ has the same sign as z, so that

$$\partial U/\partial z = \lambda z$$

where λ is a positive analytic function of x, y, z. The integral of energy may be written

$$\frac{1}{2}(x'^2 + y'^2 + z'^2) + U = 0,$$

provided that we absorb a suitable constant into U. Thus we restrict attention to the totality of motions satisfying this last relation, thereby reducing the system from the sixth to the fifth order. We shall consider only the case in which the surface $U = 0$ in space constitutes a closed simply-connected surface intersecting $z = 0$ in an oval, with $U < 0$ within the surface. The particle is then necessarily restricted to lie in the region $U \leqq 0$.

The five-dimensional manifold M of states of motion consists of the sets x, y, z, x', y', z', subject to the integral relation. The selected surface of section S will then be the four-dimensional part of M for which z vanishes with $z' = dz/dt \geqq 0$. The three-dimensional boundary $z = z' = 0$

of S is evidently made up of stream lines, since a point $z = z' = 0$ for any value t_0 of t remains always of this type.

Now the differential equation

$$\frac{d^2 z}{d t^2} + \lambda z = 0$$

shows at once that z vanishes at least once in an interval of time τ sufficiently large, but cannot vanish twice in an arbitrarily small interval of time. Hence a point of S followed along its stream line in the manifold of states of motion will cut S again within an interval 2τ of time, and always in the same sense, since for $z = 0$ we have $dz/dt > 0$ within S.

Evidently there is thus set up a one-to-one, analytic transformation of points within S into themselves. Furthermore for z, z' small, we have nearly

$$\frac{d^2 z}{d t^2} + \lambda (x, y, 0)\, z = 0,$$

where x, y are the coördinates of the motion in the plane near the given motion. But the particular solution of this equation for which $z = 0$ at $x = x_0$, $y = y_0$, $x = x_0'$, $y = y_0'$, with

$$\frac{1}{2}(x_0'^2 + y_0'^2) + U(x_0, y_0, 0) = 0$$

of course, vanishes at x_1, y_1, x_1', y_1', with

$$\frac{1}{2}(x_1'^2 + y_1'^2) + U(x_1, y_1, 0) = 0,$$

where x_1, y_1, x_1', y_1' are evidently analytic in x_0, y_0, x_0', y_0'. It is thus seen that the transformation T of S into itself can be regarded as one-to-one and continuous even along the boundary of S, provided we define T as taking the point

$$x_0,\ y_0,\ 0,\ x_0',\ y_0',\ 0$$

of S into the point

$$x_1,\ y_1,\ 0,\ x_1',\ y_1';\ 0.$$

We propose to show by means of this reduction to a transformation problem that there always exists a periodic motion intersecting the plane $z=0$ twice, unless there is a multiple periodic motion lying in the plane $z=0$.

In order to do this, let us consider the connectivity of the surface of section S. Evidently we may change the variables x, y to $\bar{x}, \bar{y}$ so that the oval $z=0,\ U\leqq 0$ becomes the circle

$$\bar{x}^2+\bar{y}^2=1.$$

If then we write

$$U=p(\bar{x},\bar{y})\,(\bar{x}^2+\bar{y}^2-1)$$

with $p>0$ within this circle, and if we write further

$$x'=\sqrt{p}\,\bar{x}',\quad y'=\sqrt{p}\,\bar{y}',\quad z'=\sqrt{p}\,\bar{z}',$$

the equation for S takes the form

$$\bar{x}'^2+\bar{y}'^2+\bar{z}'^2+\bar{x}^2+\bar{y}^2=1 \qquad (\bar{z}'\geqq 0)$$

which may be written

$$\bar{z}'=(1-\bar{x}^2-\bar{y}^2-\bar{x}'^2-\bar{y}'^2)^{1/2}.$$

Hence the interior and boundary of S are in one-to-one, continuous correspondence with the interior and boundary of the four-dimensional hypersphere

$$\bar{x}^2+\bar{y}^2+\bar{x}'^2+\bar{y}'^2=1.$$

The transformation T defines a one-to-one, continuous, direct transformation of this hypersphere into itself.

But, by a well-known theorem due to Brouwer, such a transformation leaves some point invariant. In the problem at hand, we conclude that a periodic motion exists which intersects $z = 0$ twice, (case of an interior invariant point), or else a periodic motion $z = 0$ exists (case of an invariant boundary point). But this last case is that in which the equations $x_1 = x_0$, $y_1 = y_0$, $x'_1 = x'_0$, $y'_1 = y'_0$ obtain. This clearly means that the equations of variation possess a periodic solution along this plane periodic motion in which the z component is not 0. Hence the periodic motion is multiple, and, in a certain sense there is still a periodic motion in the infinitesimal vicinity of $z = 0$, intersecting $z = 0$ twice. It seems highly probable that an actual periodic motion intersecting $z = 0$ twice must exist in all cases.

CHAPTER VI

APPLICATION OF POINCARÉ'S GEOMETRIC THEOREM

1. Periodic motions near generalized equilibrium ($m = 1$). Poincaré's last geometric theorem and modifications thereof* yield an additional instrument for establishing the existence of periodic motions. Up to the present time no proper generalization of this theorem to higher dimensions has been found, so that its application remains limited to dynamical systems with two degrees of freedom. It is our aim in this chapter to give some of the fundamental ideas involved in the theorem and its application.

It will be remembered that motion near to a periodic motion of a Hamiltonian or Pfaffian system, with m degrees of freedom and not involving the time explicitly, can be reduced to that of a similar system with only $m-1$ degrees of freedom but with an independent variable involved of period 2π. Here the periodic motion itself appears as generalized equilibrium. This reduction is accomplished by means of an analytic device (chapter IV, section 1).

In the present section we shall take up the question of the existence of motions with period $2k\pi$ near the position of generalized equilibrium for a single degree of freedom. We shall prove the existence of infinitely many such nearby periodic motions in the general stable case by a process of reasoning which, while not employing Poincaré's geometric theorem explicitly, is precisely that which establishes the theorem in certain simple cases. Later (section 3) these results are interpreted with reference to the original dynamical problem with two degrees of freedom.

* See my paper, *An Extension of Poincaré's Last Geometric Theorem*, Acta Mathematica, vol. 47 (1926).

Let the single pair of variables be p, q, so that the Hamiltonian function H involves p, q, t, being periodic in t of period 2π, and vanishes at the origin $p = q = 0$, together with its first partial derivatives, for all values of t.

If then (p_0, q_0) is any point near to the origin, there is a unique solution

$$p = f(p_0, q_0, t), \qquad q = g(p_0, q_0, t)$$

which for $t = 0$ takes on the values p_0, q_0, and which is analytic in p_0, q_0, t for t arbitrarily large and p_0, q_0 sufficiently small. Let p_1, q_1 denote the values of p, q after a complete period 2π. Evidently we have

$$p_1 = f(p_0, q_0, 2\pi), \qquad q_1 = g(p_0, q_0, 2\pi)$$

where f and g are analytic in p_0, q_0, and vanish with these variables.

In this way a transformation T is defined, of the same nature as the transformation of the surface of section obtained in the preceding chapter (section 10). For if we write $r = t$, the pair of Hamiltonian equations may be replaced by the equivalent set

$$\frac{dp}{dt} = -\frac{\partial H}{\partial q}, \qquad \frac{dq}{dt} = \frac{\partial H}{\partial p}, \qquad \frac{dr}{dt} = 1,$$

in which H is a function of p, q, r of period 2π in r. Here the manifold of states of motion is the three-dimensional p, q, r space in which the r axis represents a periodic motion, namely that corresponding to generalized equilibrium; it must not be forgotten that r is an angular variable. Now $\varphi = r = 0$ will serve as a surface of section according to our earlier work, although here we are limited to a certain neighborhood. A point $(p_0, q_0, 0)$ in this surface of section is taken along its stream line to $(p_1, q_1, 2\pi)$, i. e., to $(p_1, q_1, 0)$. Thus the transformation written above is indeed a transformation T of a surface of section S which is, however, only locally defined. Such 'local surfaces of section' can of course be constructed near a periodic motion in any dynamical problem by merely taking an element of surface which intersects but is not tangent to the corresponding stream line in the manifold of states of motion.

Now the fluid motion defined by the above three equations is that of an incompressible fluid, since the divergence of the right-hand members is 0. Consequently if we follow any tube of fluid made of sections of stream lines between the parallel planes $r = 0$ and $r = 2\pi$, which will move constantly with unit velocity in the r direction, we infer that the loss of volume at one base in time Δt is nearly $\sigma_0 \Delta t$ (σ_0, area of first base), while the equal gain at the other is nearly $\sigma_1 \Delta t$ (σ_1, area of second base). By allowing Δt to approach 0 we infer that $\sigma_0 = \sigma_1$.

Since σ_0 is an arbitrary area in $r = 0$, it is clear that T must be an area-preserving transformation of the variables p_0, q_0. This important property of T corresponds to a general property of the surface transformations associated with dynamical problems.

It is necessary now to state the conditions to be imposed upon the generalized equilibrium, with the aid of which the conclusion stated may be established.

We assume in the first place that the generalized equilibrium is of general stable type, and therefore completely stable. The significance of the normal form (chapter III, section 9) is that the solution may be written

$$p = p_0\, e^{M(p_0 q_0)t} + \Phi, \qquad q = q_0\, e^{-M(p_0 q_0)t} + \Psi,$$

in properly chosen conjugate variables p, q. Here Φ, Ψ are given as convergent power series in p_0, q_0 with initial terms of arbitrarily high degree $2\mu + 1$, and with all coefficients analytic functions of t; these series converge absolutely and uniformly for any fixed range of values for t, such as $|t| \leqq 2\pi$, when p_0, q_0 are small. The function M can be taken as a polynomial of degree not more than μ in the product $p_0 q_0$, with pure imaginary constant coefficients, of the form

$$\lambda + l p_0 q_0 + \cdots + s p_0^\mu q_0^\mu$$

with λ the multiplier. By the hypothesis of stability $\lambda / \sqrt{-1}$ is not rational, and in particular is not 0.

Our second assumption is that l is not 0. In case l vanishes but some other coefficient in M is not 0, essentially the same argument would apply. Thus the only case of failure is that in which the formal series M in the complete normal form reduces to a mere constant λ. This is a highly degenerate case, and actual examples can be constructed to show that an analogous conclusion cannot then be drawn.

The normal form gives a means of studying the transformation T. The property of the transformation T necessary for our present purposes is embodied in the following lemma whose proof is deferred to the next section:

LEMMA. For $l \neq 0$, upon suitable choice of the variables p, q, the positive quantity ε may be taken arbitrarily small, and then the integer n so large that any transformation T^ν ($\nu \leqq n$) takes the circle $r \leqq \varepsilon$ about the invariant point $r = 0$ into a region within the circle of radius 2ε, while the angular rotation effected by T^n increases from $n\lambda / \sqrt{-1}$ with r along any radial line for $r \leqq \varepsilon$, being at least 2π greater for $r = \varepsilon$ than for $r = 0$.

Let r, θ be polar coördinates and let (r_n, θ_n) denote the iterate of (r, θ) under T^n, where the rectangular coördinates p, q, the radius ε, and the integer n are selected as in the lemma. For any fixed θ_0 the difference $\theta_n - \theta_0$ will then increase from $n\lambda / \sqrt{-1}$ at $r = 0$ to a quantity at least 2π greater at $r = \varepsilon$. Hence there will be a unique solution of the equation

$$\theta_n - \theta_0 = 2k\pi$$

along a fixed radius vector, where $2k\pi$ is the least integral multiple of 2π exceeding $n\lambda / \sqrt{-1}$. Hence the analytic curve C given by this equation is met once and only once by each radius sector.

Now let us consider the image C_n of this curve under the transformation T^n; the curve C_n intersects C at some point Q, since if C_n is wholly within C or outside C, T^n would not be area-preserving in the original variables. The point Q is obtained from some point P, also on C, by the transformation T^n. Moreover P and Q have the same θ, by definition of C. Thus P and Q must coincide, and P is invariant under T^n. Since ε is arbitrarily small, we obtain the result stated:

In the case of generalized equilibrium of general stable type for a Hamiltonian problem with one degree of freedom ($l \neq 0$), *there exist infinitely many periodic motions in the vicinity.*

2. Proof of the lemma of section 1. Let us define $F(u)$ by the equation

$$l\,u F^2(u) = M(u) - \lambda.$$

It is clear that $F(u)$ is the square root of a real polynomial of degree $\mu - 1$ and constant term 1. If then we write further

$$\bar{p} = F(pq)p, \qquad \bar{q} = F(pq)q,$$

it is found that the above normal form for T is further simplified and may be written

$$\bar{p} = \bar{p}_0 e^{(\lambda + l\bar{p}_0\bar{q}_0)t} + \Phi, \qquad \bar{q} = \bar{q}_0 e^{-(\lambda + l\bar{p}_0\bar{q}_0)t} + \Psi,$$

so that all of the terms of M except the first two disappear. When the associated real variables are introduced, and we let σ and s denote the real constants $2\pi\lambda/\sqrt{-1}$ and $2\pi l/\sqrt{-1}$ respectively, we obtain formulas defining T,

$$(1) \quad \begin{aligned} p_1 &= p_0 \cos(\sigma + s r_0^2) - q_0 \sin(\sigma + s r_0^2) + P \quad (r_0^2 = p_0^2 + q_0^2), \\ q_1 &= p_0 \sin(\sigma + s r_0^2) + q_0 \cos(\sigma + s r_0^2) + Q, \end{aligned}$$

where P, Q are real power series in p_0, q_0 with initial terms of degree $2\mu + 1$ at least. It is apparent then that with these variables T is an ordinary rotation through a variable angle $\sigma + s r_0^2$, except for terms of order $2\mu + 1$ in the distance from the origin. It is such a choice of variables that will be adopted. If $l \neq 0$, we may take s as positive.

Hence we have, in some fixed neighborhood of the origin and for a fixed K,

$$|P|,\ |Q| \leqq K r_0^{2\mu+1}. \tag{2}$$

From the above formulas we find at once

$$r_1^2 = r_0^2 + R, \tag{3}$$

where the series R begins with terms of at least the degree $2\mu+2$. This gives, for the increment $\Delta r_0^2 = r_1^2 - r_0^2$,

$$|\Delta r_0^2| \leqq L r_0^{2\mu+2} \tag{4}$$

where L is a fixed constant.

From (4) it is seen that r_n^2 increases less rapidly upon successive iteration than if

$$d r_n^2 / dn = L r_n^{2\mu+2}$$

But this yields

$$r_n = r_0 / (1 - L\mu r_0^{2\mu} n)^{1/2\mu}.$$

Hence r_n can only increase to twice the initial value r_0 after $n \geqq \nu$ iterations, where

$$L\mu\nu = (1 - 2^{-2\mu}) / r_0^{2\mu},$$

that is for n of at least the order of $r_0^{-2\mu}$. Likewise r_n can only decrease to half the initial value for n of the same order. These results may be combined in the form

$$\frac{1}{2} \leqq r_n / r_0 \leqq 2 \qquad (n \leqq N r_0^{-2\mu}). \tag{5}$$

This is our first important conclusion.

Likewise since we have from (1)

$$\frac{q_1 p_0 - p_1 q_0}{p_1 p_0 + q_1 q_0} = \frac{(p_0^2 + q_0^2)\sin(\sigma + s r_0^2) + p_0 Q - q_0 P}{(p_0^2 + q_0^2)\cos(\sigma + s r_0^2) + q_0 Q + p_0 P},$$

it follows that we have

$$(6) \qquad \theta_1 = \theta_0 + \sigma + s r_0^2 + \Theta$$

where Θ is a power series in r_0 beginning with terms of degree 2μ at least, with coefficients of simple trigonometric type in θ_0. Moreover Θ is uniformly and absolutely convergent for r_0 sufficiently small, and its partial derivatives are given by derived series of a similar sort. This formula shows that for $r_0 = 0$, we have $\theta_n = \theta_0 + n\sigma$, while for $r_0 > 0$, the difference $\theta_n - \theta_0 - n\sigma$ can be made arbitrarily large by taking n sufficiently large but in the range (5).

By the aid of (3) and (6) we obtain

$$(7) \qquad \begin{cases} \left|\dfrac{\partial \varrho_1}{\partial \varrho_0} - 1\right| \leqq A\varrho_0^{\mu}, & \left|\dfrac{\partial \varrho_1}{\partial \theta_0}\right| \leqq A\varrho_0^{\mu+1}, \\ \left|\dfrac{\partial \theta_1}{\partial \varrho_0} - s\right| \leqq A\varrho_0^{\mu-1}, & \left|\dfrac{\partial \theta_1}{\partial \theta_0} - 1\right| \leqq A\varrho_0^{\mu}, \end{cases}$$

where we have written $\varrho = r^2$, and where A is a suitably chosen positive constant.

But the identities

$$\frac{\partial \varrho_n}{\partial \varrho_0} = \frac{\partial \varrho_n}{\partial \varrho_{n-1}} \frac{\partial \varrho_{n-1}}{\partial \varrho_0} + \frac{\partial \varrho_n}{\partial \theta_{n-1}} \frac{\partial \theta_{n-1}}{\partial \varrho_0},$$

$$\frac{\partial \theta_n}{\partial \varrho_0} = \frac{\partial \theta_n}{\partial \varrho_{n-1}} \frac{\partial \varrho_{n-1}}{\partial \varrho_0} + \frac{\partial \theta_n}{\partial \theta_{n-1}} \frac{\partial \theta_{n-1}}{\partial \varrho_0},$$

may be written

$$(8) \qquad \begin{aligned} u_n &= (1 + \varepsilon_1) u_{n-1} + \varepsilon_2 v_{n-1}, \\ v_n &= (s + \varepsilon_3) u_{n-1} + (1 + \varepsilon_4) v_{n-1}, \end{aligned}$$

in which we put

$$u_n = \partial \varrho_n / \partial \varrho_0, \qquad v_n = \partial \theta_n / \partial \varrho_0,$$

while, by (5) and (7), for $n \leqq N\varrho_0^{-\mu}$,

$$|\varepsilon_1| \leqq 2^{\mu} A\varrho_0^{\mu}, \qquad |\varepsilon_2| \leqq 2^{\mu+1} A\varrho_0^{\mu+1}, \qquad |\varepsilon_3| \leqq 2^{\mu-1} A\varrho_0^{\mu-}$$

$$|\varepsilon_4| \leqq 2^{\mu} A\varrho_0^{\mu}.$$

These equations (8) enable us to determine u_n, v_n in succession for $n = 1, 2, \cdots$, with the initial conditions $u_0 = 1$, $v_0 = 0$.

Suppose now for a moment that the small terms be neglected in these equations (8). They then take the form

$$u_n = u_{n-1}, \qquad v_n = s u_{n-1} + v_{n-1},$$

whence, by elimination of u, we obtain

$$\Delta^2 v_n = v_{n+2} - 2 v_{n+1} + v_n = 0.$$

It is easily verified that the complete result of elimination yields similarly

$$\Delta^2 v_n = \varepsilon_5 \Delta v_n + \varepsilon_6 v_n, \tag{9}$$

in which we have

$$|\varepsilon_5|, |\varepsilon_6| \leqq B \varrho_0^{\mu-1} \qquad (n \leqq N \varrho_0^{-\mu}) \tag{10}$$

within a small region about the origin. Furthermore, the initial conditions may be written

$$v_0 = 0, \qquad \Delta v_0 = s + \varepsilon_3^0, \tag{11}$$

where ε_3^0 denotes the value of ε_3 when $\varrho_{n-1}, \theta_{n-1}$ are replaced by ϱ_0, θ_0 respectively.

It is obvious that $v_1 = s + \varepsilon_3^0$ is positive and thence, by use of (9), $v_2, v_3, \cdots$ are also positive for $n = 1, 2, \cdots$ until n becomes large if ϱ_0 is small enough, the approximate value of v_n being ns. We desire to obtain a more definite idea of the range of values for ϱ_0 and n for which v_n remains positive. During this range the angular variable θ_n increases with r_0, for a fixed angle θ_0.

Now the equation (9) is a homogeneous linear difference equation of the second order in v_n, and we are considering the particular solution satisfying (11). Evidently v_n will remain positive so long as Δv_n continues positive. A first question is then to determine the range of values of n throughout which both v_n and Δv_n necessarily remain positive. But the linear difference equation yields

$$\Delta^2 v_n \geqq -B\varrho_0^{\mu-1}[|\Delta v_n|+|v_n|],$$

so that clearly v_n and Δv_n diminish less rapidly than if

$$\Delta^2 v_n = -B\varrho_0^{\mu-1}[\Delta v_n + v_n],$$

while v_n and Δv_n remain positive. Thus v_n and Δv_n will remain positive for $n = 1, 2, \cdots$, at least as long as for the solution of the linear difference equation of the second order with constant coefficients

$$v_{n+2} - (2 - B\varrho_0^{\mu-1})\, v_{n+1} + v_n = 0$$

satisfying the initial conditions (11). But this solution is

$$v_n = (s+\varepsilon_3^0)\frac{e^{\alpha_1 n} - e^{\alpha_2 n}}{e^{\alpha_1} - e^{\alpha_2}} \tag{12}$$

where α_1, α_2 are defined by the equation

$$\alpha_i = \log\left(1 - \frac{1}{2}B\varrho_0^{\mu-1} \pm \sqrt{-\left(B\varrho_0^{\mu-1} - \frac{1}{4}B^2\varrho_0^{2\mu-2}\right)}\right) \qquad (i = 1, 2).$$

But v_n and Δv_n as thus determined will certainly remain positive until dv_n/dn vanishes, i. e.

$$e^{(\alpha_1-\alpha_2)n} = \alpha_2/\alpha_1.$$

Since the leading term in $\alpha_1 - \alpha_2$ is clearly

$$2\sqrt{-B}\,\varrho_0^{(\mu-1)/2}$$

while the leading term in α_2/α_1 is -1, this relationship shows that n must be of the reciprocal order $\varrho_0^{-(\mu-1)/2}$.

Hence we infer that so long as $n \leqq N^* r_0^{-(\mu-1)}$ (compare with (5)), the angle θ_n will increase with r_0 for fixed θ_0 in the prescribed neighborhood.

The nature of the inequalities derived above makes it clear that we can select a value of r_0 so small, and then of n so large, that the conditions laid down in the lemma are satisfied.

3. Periodic motions near a periodic motion ($m = 2$). We have already seen (chapter IV, section 1) how the general Pfaffian system in which the time t does not appear explicitly admits of a reduction to a similar system with one less degree of freedom, provided that we are considering motions near a given periodic motion. In the reduced equations, however, an angular variable of period 2π appears in the differential equations, and the given periodic motion takes the form of generalized equilibrium.

In this section we propose to consider the periodic motions near a given periodic motion for the special Hamiltonian case ($m = 2$)

$$(13) \qquad \frac{dp_i}{dt} = -\frac{\partial H}{\partial q_i}, \quad \frac{dq_i}{dt} = \frac{\partial H}{\partial p_i} \quad (i = 1, 2),$$

in which H is an analytic function of p_1, q_1, p_2, q_2, not involving t. However such a periodic motion admits of analytic continuation with variation of the energy constant $H = h$ (chapter V, section 9), and so is not isolated. Our aim then will be to consider only those nearly periodic motions which belong to the same value of h as the given periodic motion; this value may be taken to be $h = 0$.

The possibility of reduction to a Pfaffian case $m = 1$, combined with the results of the preceding section renders it highly probable at the outset that there will in general be infinitely many nearby periodic motions of long period, provided that the given periodic motion is of stable type.

In considering this question, we shall make the further assumption that the given Hamiltonian problem is associated with an ordinary Lagrangian problem whose principal function L

is quadratic in the velocities. If q_1, q_2 are the coördinates in this Lagrangian problem, the equations

$$p_i = \frac{\partial L}{\partial q_i'} \qquad (i = 1, 2)$$

serve of course to define the variables p_1, p_2.

Let $q_1 = q_1(t)$, $q_2 = q_2(t)$ be the equations yielding this periodic motion of period t, and consider the corresponding analytic curve in the q_1, q_2 plane.

Evidently we can introduce a modified system of coördinates $\overline{q}_1, \overline{q}_2$ such that $\overline{q}_2$ vanishes along the motion, while $\overline{q}_1$ increases by 2π as a point makes a circuit of the motion. For instance, if the curve of motion is without double points, it may be deformed into a circle concentric with the origin, in which case $\overline{q}_1$ and $\overline{q}_2$ may be taken as angle and radial displacement respectively. It is clear indeed that we may take $\overline{q}_1 = 2\pi t/\tau$ along the periodic motion. Of course such a change of variables from q_1, q_2 to $\overline{q}_1, \overline{q}_2$ does not affect the Lagrangian character of the dynamical problem, although the new principal function L is periodic of period 2π in the variable $\overline{q}_1$.

The corresponding Hamiltonian problem will have the form (13) in which H is periodic of period 2π in q_1, while for the periodic motion under consideration we have $q_1 = 2\pi t/\tau$, $q_2 = 0$. From the Hamiltonian equations in these variables, we have also along the periodic motion

$$2\pi/\tau = \partial H/\partial p_1, \qquad 0 = \partial H/\partial p_2.$$

It is obvious then that we may solve the equation $H = h$ for p_1 in the form

$$p_1 + K(q_1, p_2, q_2, h) = 0 \tag{14}$$

where K is a real, single-valued, analytic function of its four arguments, periodic in q_1 of period 2π. Furthermore we may regard h, q_1, p_2, q_2 as the dependent variables instead of p_1, q_1, p_2, q_2; we observe that (14) may be solved explicitly for h, since from the relation $H = h$, we derive

$$\frac{\partial H}{\partial p_1} \frac{\partial p_1}{\partial h} = 1$$

so that $\partial p_1/\partial h \neq 0$ along the motion. When these variables are used instead of p_1, q_1, p_2, q_2, the variational principle (chapter II, section 10) takes the form

$$\delta \int_{t_0}^{t_1} (-K q_1' + p_2 q_2' - h)\, dt = 0, \tag{15}$$

which leads to the four equations

$$\frac{\partial K}{\partial h} \frac{dq_1}{dt} + 1 = 0, \quad \frac{\partial K}{\partial h} \frac{dh}{dt} + \frac{\partial K}{\partial p_2} \frac{dp_2}{dt} + \frac{\partial K}{\partial q_2} \frac{dq_2}{dt} = 0,$$

$$\frac{dq_2}{dt} = \frac{\partial K}{\partial p_2} \frac{dq_1}{dt}, \quad \frac{dp_2}{dt} + \frac{\partial K}{\partial q_2} \frac{dq_1}{dt} = 0.$$

From these equations we infer directly $h = \text{const.}$, which we know to be true of course.

Now it is evident that near the given periodic motion q_1 can serve as independent variable as well as t. If we eliminate t in the above equations, we find

$$\frac{dp_2}{dq_1} = -\frac{\partial K}{\partial q_2}, \quad \frac{dq_2}{dq_1} = \frac{\partial K}{\partial p_2}. \tag{16}$$

Here we are to set $h = 0$ in the function K, and K is periodic of period 2π in q_1. The given periodic motion corresponds to

$$p_2 = \varphi(q_1), \quad q_2 = 0$$

where φ is periodic of period 2π in q_1. These equations are clearly in Hamiltonian form ($m = 1$), with a generalized equilibrium point at the origin, at least after the simple modification

$$\bar{p}_2 = p_2 - \varphi(q_1), \ \bar{q}_2 = q_2,$$

in which we may take

$$\bar{K} = K - \varphi'(q_1)q_2.$$

Conversely, if we have a solution of (16), we can determine t from the equation

$$\frac{dt}{dq_1} = -\frac{\partial K}{\partial h},$$

and obtain a solution of the original system when t is taken as independent variable. Thus (13) and (16) are equivalent.* Periodic motions near the given periodic motion for (13) correspond to motions near the origin of period $2k\pi$ for (16).

For a Hamiltonian problem (13) *which reduces to a generalized equilibrium problem* (16) *of stable type* ($l \neq 0$), *there will exist infinitely many periodic motions in the vicinity of the given periodic motion, making in general many circuits of that motion before re-entering.*

This result is of course obtained as the direct application of section 1 for the reduced problem.

4. Some remarks. The general conclusion which appears in consequence of the preceding sections is that, for a given value of the energy constant, there exist *in general* periodic motions in the vicinity of a periodic motion of stable type, at least when the dynamical system has two degrees of freedom and is of ordinary type. The fact that there may exist isolated periodic motions of stable type, even for dynamical systems with two degrees of freedom, may be brought out by means of the following elementary example.

Let us write

$$H = \frac{1}{2}k^2(p_1^2 + q_1^2) + \frac{1}{2}l^2(p_2^2 + q_2^2),$$

where the quantities k, l are incommensurable with one another, of which the general solution is

* For the reduction employed, cf. Whittaker, *Analytical Dynamics*, chap. 12.

$$p_1 = A\cos kt + B\sin kt, \quad q_1 = -A\sin kt + B\cos kt,$$
$$p_2 = C\cos lt + D\sin lt, \quad q_2 = -C\sin lt + D\cos lt.$$

The energy constant h is defined by the relation

$$H = \frac{1}{2}k^2(A^2+B^2) + \frac{1}{2}l^2(C^2+D^2) = h.$$

The only periodic solutions are the two analytic families

$$p_1 = q_1 = 0 \text{ and } p_2 = q_2 = 0,$$

all of which are of stable type. For an assigned value of the energy constant, there are only two such periodic motions; thus all periodic motions of the second family with assigned A^2+B^2 represent the same closed curve in the three-dimensional manifold $H = h$ in four-dimensional p_1, q_1, p_2, q_2 space. If the transformation T be set up for this case as in the preceding section 2, it is found to be essentially a rotation through an angle incommensurable with 2π, and so to correspond precisely to the highly degenerate case there excluded from consideration.

A first question as to a possible generalization of the above results in the case $m = 2$ is the following: Suppose the origin is a point of generalized equilibrium of general stable type for a given differential system which is, in addition, completely stable; if the constant l is not 0, does it follow that there will always exist infinitely many periodic motions in the vicinity of the origin?

It seems to me very doubtful that the answer is in the affirmative. In the preceding argument the area-preserving property played a vital part. For this more general completely stable type, there is no reason to believe that this property continues to hold, even in the Pfaffian case.

The example can be generalized so as to indicate a preliminary necessary requirement if the conclusion that there are infinitely many periodic motions near a given stable periodic motion is to hold for Hamiltonian systems with more than two degrees of freedom.

In fact, consider the case of a dynamical system

$$\frac{dp_i}{dt} = -\frac{\partial H}{\partial q_i}, \quad \frac{dq_i}{dt} = \frac{\partial H}{\partial p_i} \qquad (i = 1, \cdots, m)$$

in which H is a function of the m products $p_1 q_1, \cdots, p_m q_m$ with p_i, q_i conjugate imaginary variables, namely

$$H = \sum_{j=1}^{m} c_j p_j q_j + \frac{1}{2} \sum_{j,k=1}^{m} d_{jk} p_j q_j p_k q_k.$$

Here the coefficients c_i, d_{ij} are periodic of period τ in t. The origin is a point of generalized equilibrium with multipliers

$$\lambda_i = \int_0^\tau c_i(t)\, dt \qquad (i = 1, \cdots, m),$$

so that it will be of general stable type if these m quantities and $2\pi\sqrt{-1}/\tau$ have no linear commensurability relations. If for the sake of brevity we write

$$x_i = \int_0^t \left[c_i + \sum_{j=1}^{m} d_{ij} p_j^0 q_j^0 \right] dt,$$

the general solution is at once found to be

$$p_i = p_i^0 e^{-x_i}, \quad q_i = q_i^0 e^{x_i} \quad (i = 1, \cdots, m).$$

Moreover if we write

$$C_i = \int_0^\tau c_i\, dt, \quad D_{ij} = \int_0^\tau d_{ij}\, dt,$$

the condition that the solution is periodic of period $k\tau$ is

$$C_i + \sum_{j=1}^{m} D_{ij} p_j^0 q_j^0 = 2\pi k_i \sqrt{-1}/k \qquad (i = 1, \cdots, m),$$

where $k_1, \cdots, k_m$ are integers. But these form m linear, non-homogeneous, algebraic equations in $p_i^0 q_i^0$, $(i = 1, \cdots, m)$, which can be solved if the determinant $|D_{ij}|$ is not 0. Moreover by making the ratios k_i/k small, the periodic motion can be taken near the origin. On the other hand if $|D_{ij}| = 0$, such a determination will be impossible in general.

In the particular case when the c_i's and d_{ij}'s are constants, the system appears in complete normal form, and the c_i's are the multipliers, while the d_{ij}'s are invariants analogous to l in the case $m = 1$.

Hence at least the condition $|d_{ij}| \neq 0$ must be imposed in the case $m > 1$, as analogous to the condition $l \neq 0$ in the case $m = 1$, if an infinitude of nearly periodic motions is to be anticipated in all cases.

Any generalization must of course take proper account of the uniform analytic integrals (such as the energy integral) which exist. In fact, if there are k of these integrals which are independent, the given stable periodic motion will admit of k-fold analytic continuation. Evidently it is not such periodic motions of the same analytic family as the given motion which interest us, but rather nearby periodic motions for the same values of the constants of integration as the given periodic motion, and making many circuits of it in a period.

5. The geometric theorem of Poincaré.* Poincaré showed that the existence of an infinite number of periodic orbits in the restricted problem of three bodies and other dynamical problems would follow at once from a certain geometric theorem to which the lemma of section 1 is intimately related.

For convenience we shall first state:

POINCARÉ'S THEOREM. Given a ring $0 < a \leq r \leq b$ in the r, θ plane (r, θ being polar coördinates), and a one-to-one, continuous, area-preserving transformation T of the ring into itself, which advances points on $r = a$ and regresses points on $r = b$. Then there will exist at least two points of the ring invariant under T.

* This section is essentially the same as section 34 of my paper, *Dynamical Systems with Two Degrees of Freedom*, Trans. Amer. Math. Soc., vol. 18 (1917).

We will indicate briefly the proof of this theorem.

Let us take $x = \theta$, $y = r^2$ as the rectangular coördinates of a point in the x, y plane. The ring then appears as a strip $a^2 \leqq y \leqq b^2$. The transformation T of this strip advances points of the boundary $y = a^2$ to the right, and moves points on $y = b^2$ to the left. Moreover T is area-preserving in the x, y plane (for we have $2\,r\,d\,r\,d\,\theta = d\,x\,d\,y$), and displaces any two points which have the same ordinate and whose abscissas differ by a multiple of 2π in the same way.

Let us combine T with a further transformation T_ε which effects a translation of the x, y plane in the direction of the y axis through a distance $\varepsilon > 0$. The transformation T followed by T_ε yields an area-preserving transformation $T\,T_\varepsilon$ which shifts the given strip into the strip $a^2 + \varepsilon \leqq y \leqq b^2 + \varepsilon$.

Suppose if possible that there exists no invariant point of T. There exists then a positive quantity d such that all points are displaced at least a distance d by the transformation T. Choose ε less than d.

Consider now the narrow strip $a^2 \leqq y \leqq a^2 + \varepsilon$. By the transformation $T\,T_\varepsilon$ the lower edge of this strip is carried into the upper edge and the strip is carried into a second strip lying wholly above the first one save along the common edge. By a repetition of the transformation $T\,T_\varepsilon$ the second strip goes into a third, and so on.

By a continuation of this process, a series of strips is obtained forming consecutive strata. Each of these strata is unaltered by a shift of 2π to the right. This follows from the fact that T and T_ε are single-valued over the ring.

The images of these strata on the ring are a set of closed strata about the ring, all having equal area of course since $T\,T_\varepsilon$ is an area-preserving transformation in the r, θ as well as in the x, y plane. Consequently some one of the strata on the infinite strip, say the k-th, must overlap the upper edge $y = b^2$.

In the x, y plane let Q be a point of the upper edge of the k-th stratum for which y is a maximum. Let P be the point of $y = a^2$ from which Q is derived by k-fold repetition of TT_ε, and let P', P'', $\cdots$, $P^{(k)} = Q$ denote the successive images of P under the iteration of TT_ε. Draw the straight line PP' which will obviously lie on the first stratum. The successive images of this line, PP', $P'P''$, $\cdots$, $P^{(k-1)}P^{(k)}$ will lie in the successive strata, and will have no points in common except that successive arcs have an end-point in common. Thus we get a single arc PQ, made up of all these lines, which is without double points.

Consider now a vector LL' drawn from a point L to its image L' under TT_ε, of which the initial point moves from

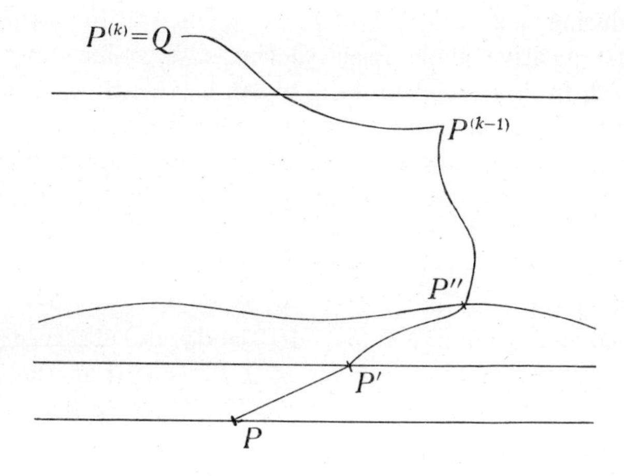

P to $P^{(k-1)}$ along the line PQ. The angle which this vector makes with the positive direction of the x axis at the outset may be taken to be a positive acute angle, since the image P' of P lies to the right of and above P. When L has varied to its final position $P^{(k-1)}$, the same angle lies in the second or third quadrant, since $P^{(k)}$ lies to the left of $P^{(k-1)}$, by the hypothesis of the theorem.

Our construction of the successive arcs PP', $P'P''$, $\cdots$, renders it apparent that as L moves from P to $P^{(k-1)}$, its image L' moves along the same curve from P' to Q. Therefore we see at once from the figure that LL' has rotated through the least positive angle from the first direction to the second. If L' be moved further in a vertical direction from Q up to $y = b^2 + \varepsilon$ the same statement will still hold of the standard curve described by L provided that ε is small, since Q lies at most ε above $y = b^2$.

Suppose now that L moves in any manner from a point of $y = a^2$ to a point of $y = b^2$ in the given strip. The transformation TT_ε leaves no points of this region invariant, so that the point L will never coincide with L'. In the initial position for L on $y = a^2$ the angle made by LL' lies in the first quadrant. In the final position it lies in the second or third quadrant. But the total variation of angle during the variation of L has been seen to be through the least positive angle in a special case. Since any one path of L from $y = a^2$ to $y = b^2$ can be varied continuously into any other, the same must be true always.

Now let ε approach zero. As ε becomes smaller the vector LL' continues to have a definite direction, since no invariant points under TT_ε are present. By a limiting process we infer that, for the transformation T, the angular variation of LL' is through the least positive angle consistent with its initial and final directions. It should be observed that for L on $y = a^2$ the direction of LL' is that of the positive x axis, while on $y = b^2$ the direction is that of the negative x axis.

Consider now the inverse transformation T^{-1} which is of the same type as T although it moves points on $y = a^2$ to the left, and points on $y = b^2$ to the right. By an entirely analogous argument to that given above we are led to infer that if a vector $LL^{(-1)}$ with end-point $L^{(-1)} = T^{-1}(L)$ has its initial point L varied from a point of $y = a^2$ to a point of $y = b^2$, the total angular variation will be the least negative angle consistent with its initial and final positions.

But the total rotation of $LL^{(-1)}$ is precisely the same as that of the oppositely directed vector $L^{(-1)}L$ which joins a point $L^{(-1)}$ of $y = a^2$ to its image L under T.

Hence by our earlier result the total angular variation of $L^{(-1)}L$ must also be the least positive angle consistent with the two positions. Thus we have been led to a contradiction, so that there must exist at least one invariant point.

To prove that there are at least two invariant points we may adopt the method used by Poincaré.

Let the point L describe the fundamental rectangle

$$0 \leqq x \leqq 2\pi, \qquad a^2 \leqq y \leqq b^2,$$

in the x, y plane, in a positive sense. It is obvious that the total rotation of the vector LL' is 0 over this circuit, since there is no rotation along $y = a^2$ or $y = b^2$, and the rotations along $x = 0$ and $x = 2\pi$ are the negatives of one another. But around a simple invariant point the rotation is $\pm 2\pi$. Hence, according to obvious reasoning, there will either be at least two simple invariant points with rotations $+2\pi$ and -2π, or there will be at least one multiple invariant point. As a matter of fact, there will always be at least two geometrically distinct invariant points.*

6. The billiard ball problem.† In order to see how the theorem of Poincaré and its generalization can be applied, we will consider first a special but highly typical system of this sort, namely that afforded by the motion of a billiard ball upon a convex billiard table. This system is very illuminating for the following reason: Any Lagrangian system with two degrees of freedom is isomorphic with the motion of a particle on a smooth surface rotating uniformly about a fixed axis and carrying a conservative field of force with it.‡ In particular if the surface is not rotating and if the field of force is lacking, the paths of the particle will be geodesics. If the surface is now flattened to the form of a plane convex curve C, the 'billiard ball problem' results.

* See my paper, *An Extension of Poincaré's Last Geometric Theorem*, Acta Mathematica, vol. 47 (1926).

† The sections 6–9 are taken from my paper *On the Periodic Motions of Dynamical Systems*, about to appear in the Acta Mathematica.

‡ See my paper, *Dynamical Systems with Two Degrees of Freedom*, Trans. Amer. Math. Soc., vol. 18 (1917), section 7. It is assumed that the Lagrangian principal function is quadratic in the velocities.

But in this problem the formal side, usually so formidable in dynamics, almost completely disappears, and only the interesting qualitative questions need to be considered. If C is an ellipse an integrable problem results, namely the limiting case of an ellipsoid treated by Jacobi.

In the billiard ball problem one can arrive at the existence of certain periodic motions by direct maximum-minimum methods. As of interest in itself I wish to show how this can be done. Results which are being obtained by Morse (see chapter V, section 8) indicate that the scope of these methods, already developed to some extent by Poincaré, Hadamard, Whittaker and myself, can be further extended. Thus the power of such maximum-minimum considerations in the billiard ball problem is likely to prove typical of the general case.

The longest chord of the boundary C of the billiard table, when traversed in both directions, evidently yields one of the simplest periodic motions. The billiard ball moving along this chord strikes the curved boundary at right angles and recoils along it in the opposite direction. If we seek to vary this chord continuously, while diminishing its length as little as possible, so as finally to interchange its two ends, there will be an intermediate position of least length, which will be the chord crossing C where C is of least breadth. Detailed computation of the slightly perturbed motions indicates that the first of these two periodic motions is unstable, while the second may be stable, or unstable.

Next we ask for the triangle of maximum length inscribed in C. Evidently at least one such triangle will exist, and can have no degenerate side of length 0. At each of its vertices the tangent will of course make equal angles with the two sides passing through the vertex. Hence a 'harmonic triangle' is obtained which will correspond to two distinct motions, one for each of the two possible senses of description.

Moreover, if we seek to vary this triangle continuously, not changing the order of its vertices and diminishing the perimeter as little as possible, so as finally to advance the vertices cyclically, we discover a second harmonic triangle, also corresponding to two periodic motions.

In this way the existence of two harmonic n-sided polygons which make $k \leqq n/2$ circuits of the curve C (k prime to n) can be proved. The two periodic motions corresponding to the polygon of maximum type will be unstable, while the others of minimax type may be stable or unstable.

In the case of a circular boundary the totality of regular inscribed polygons (simple or cross) form the harmonic polygons.

7. The corresponding transformation T. We propose next to set up a ring transformation T associated with the billiard ball problem, and to show how the geometric theorem of Poincaré in its first form leads to the conclusion deduced above. The reduction to a ring transformation is of fundamental theoretic importance, quite aside from the relation to the question of periodic motions. It should be noted also that, in the cases of most interest like the restricted problem of three bodies, the method of reduction to a ring transformation and application of the geometric theorem of Poincaré is available for the treatment of the periodic motions, while the maximum-minimum method has not as yet been shown to be applicable.

To begin with we suppose the length of C to be 2π and to be measured from a fixed point to a variable point P by an angular coördinate φ.

At P, taken as the point of projection of the billiard ball, let θ denote the angle between the positive direction of the tangent and the direction of projection. The variable θ varies between 0 and π only. These coördinates θ, φ suffice to represent all possible states of projection unambiguously. If φ be taken as an angular coördinate in the plane, while θ, augmented by a constant, say π, be taken as a radial coördinate, the set of values (θ, φ) is represented on a ring bounded by concentric circles of radii π and 2π respectively, namely the circles $\theta = 0$ and $\theta = \pi$.

Consider now a definite state of projection at P with given θ, φ. The billiard ball leaves the table at P, to strike it again at P_1, there to be projected in a state (θ_1, φ_1), say, and so forth indefinitely. If C is an analytic curve, as we assume it to be, the correspondence between θ, φ and θ_1, φ_1 is evidently one-to-one and analytic within the ring. When θ is nearly 0 or π, the ball is projected at a slight angle to the edge, and strikes it again at a nearby point with θ nearly 0 or π as the case may be. Hence the points on the bounding circles correspond to themselves with $\theta_1 = \theta$, $\varphi_1 = \varphi$.

One further remark needs to be made about the correspondence along the two boundaries of the ring. If we think of each point (θ, φ) as being carried into (θ_1, φ_1) by a transformation or deformation of the ring, this transformation T will effect a certain number of complete rotations of the inner circle, and also of the outer circle, since the points of these boundaries are invariant as just seen. We may arbitrarily regard the inner circle as having undergone no rotation, but the same will not then be true of the outer circle, which can at once be shown to have undergone a single complete revolution in the positive sense. For let the projection angle θ, for a given point P and corresponding fixed φ, vary from 0 to π. It is obvious that then θ_1 will increase from 0 to π while φ increases by 2π since the point P_1 makes a complete circuit of P in a positive sense. In other words, the transformation T takes radial segments across the ring into curves starting at the same point of the inner circle, but winding around the ring just once while crossing it. Hence the outer boundary has undergone a single positive revolution under the transformation T.

Suppose now that we have a periodic motion, for example that corresponding to one of the harmonic triangles taken in a positive sense. It is evident that the transformation T of the ring takes the point of the ring representing the state of projection at the first vertex into that at the second; and likewise takes the state for the second vertex into that for the third, and that for the third vertex into the first. Thus when T is applied, the triple of points on the ring is cyclically advanced, and each point of the triple is unaltered by the application of the third iterate T^3 of T. Conversely, to any triple with this property, or to any point invariant under T^3, together with its images under T and T^2, corresponds a motion belonging to a harmonic triangle. Evidently then, from considerations advanced earlier, there are at least four such triples. It is obvious that there can be no invariant points under T itself because φ is increased but by less than 2π.

In this way the search for harmonic polygons and the allied periodic motions in the billiard ball problem resolves itself into the determination of sets of distinct points $P_1, \cdots, P_n$ cyclically advanced by T, so that in general $T^n(P_i) = P_i$. More generally, each and every interesting property of the motion of the billiard ball is mirrored in a corresponding property of the transformation T. Thus the dynamical problem is effectively reduced to that of a particular transformation of a circular ring into itself.

8. Area-preserving property of T. There is a further property of the transformation

$$T: \quad \theta_1 = f(\theta, \varphi), \quad \varphi_1 = g(\theta, \varphi),$$

which plays a fundamental part in applying the geometric theorem of Poincaré: the double integral $\int\int \sin\theta \, d\theta \, d\varphi$ taken over any area σ of the ring has the same value as over the image of σ under T and its iterates. This is essentially an area-preserving property in modified coördinates.

Before passing to the entirely elementary proof of this fact, one immediate application may be cited in justification of the earlier statement as to the fundamental theoretic importance of the ring transformation. Since the integral evaluated over $\sigma_1, \sigma_2 \cdots$, has the same value, and since its value over the entire ring is finite, being 4π, some two of the images σ_i and σ_j overlap. Employing the inverse transformation we infer that σ_{i-1} and σ_{j-1} also overlap, and thus finally that σ_{i-j} and σ_0 overlap $(i > j)$. But, interpreted for the billiard ball problem, this means that the ball can be projected very nearly with arbitrary position and direction to return subsequently to nearly the same position and direction. As elaborated by Poincaré,* this chain of reasoning leads to the conclusion that the 'probability' is unity that an arbitrary motion returns infinitely often to the neighborhood of its initial state. He called this property of the dynamical system 'stability in the sense of Poisson'.

The proof that the double integral is invariant depends on an explicit evaluation of the determinant

$$J = \frac{\partial \theta_1}{\partial \theta}\frac{\partial \varphi_1}{\partial \varphi} - \frac{\partial \theta_1}{\partial \varphi}\frac{\partial \varphi_1}{\partial \theta}.$$

In fact, if $\int\int M(\theta, \varphi)\, d\theta\, d\varphi$ is invariant we have

$$\int\int M(\theta_1, \varphi_1)\, d\theta_1\, d\varphi_1 = \int\int M(\theta, \varphi)\, d\theta\, d\varphi$$

where the variables θ_1, φ_1 range over the region σ_1, just as θ, φ do over σ. But according to the fundamental theorem for change of variables, the change of variables T gives the integral on the left the form

$$\int_\sigma\int M(\theta_1, \varphi_1)\, J\, d\theta\, d\varphi.$$

Comparing this expression and the integral on the right, which are both integrals over the same arbitrary region σ, we deduce the functional relation

$$M(\theta_1, \varphi_1)\, J = M(\theta, \varphi)$$

as the necessary and also sufficient condition for invariance. Hence to establish that $\int\int \sin\theta\, d\theta\, d\varphi$ is invariant, we need only prove

$$J = \sin\theta / \sin\theta_1.$$

Let

$$x = F(\varphi), \qquad y = G(\varphi)$$

* See his *Méthodes nouvelles de la Mécanique céleste,* vol. 3, chap. 26.

be the equations of C in rectangular coördinates, so that, if τ denotes the angle between the positive x axis and the positive tangential direction at a point of C, we have

$$\tau = \tan^{-1}\frac{G'(\varphi)}{F'(\varphi)}.$$

Similarly let τ_1 denote the like angle at the transformed point, which will be given by the same expression save that φ is replaced by φ_1. Finally let α designate the angle between the positively directed axis and the direction of initial projection (figure).

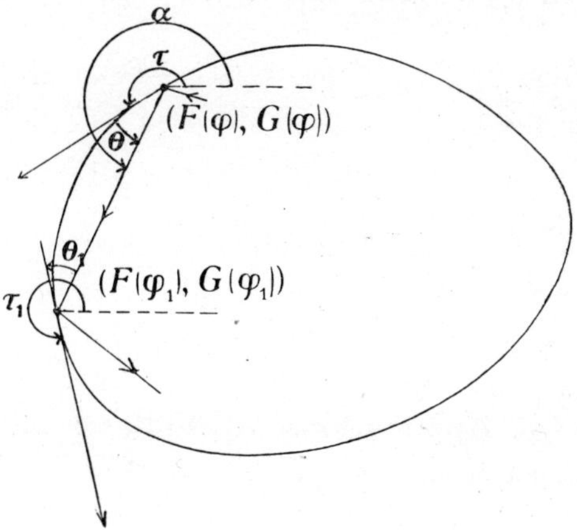

It is evident that the following two relations will hold

$$\theta = \alpha - \tau,$$
$$\theta_1 = \tau_1 - \alpha.$$

Substituting in the above value for τ and the analogous value for τ_1, and also substituting in for α the value

$$\tan^{-1}\frac{G(\varphi_1)-G(\varphi)}{F(\varphi_1)-F(\varphi)},$$

evident by inspection, we obtain the explicit formulas

$$T:\begin{cases}\theta = \tan^{-1}\dfrac{G(\varphi_1)-G(\varphi)}{F(\varphi_1)-F(\varphi)} - \tan^{-1}\dfrac{G'(\varphi)}{F'(\varphi)} = L(\varphi, \varphi_1),\\[2ex] \theta_1 = \tan^{-1}\dfrac{G'(\varphi_1)}{F'(\varphi_1)} - \tan^{-1}\dfrac{G(\varphi_1)-G(\varphi)}{F(\varphi_1)-F(\varphi)} = M(\varphi, \varphi_1).\end{cases}$$

These two equations define the transformation T from (θ, φ) to (θ_1, φ_1). Taking differentials, we find

$$d\theta = L_\varphi\, d\varphi + L_{\varphi_1}\, d\varphi_1, \qquad d\theta_1 = M_\varphi\, d\varphi + M_{\varphi_1}\, d\varphi_1,$$

whence at once

$$d\theta_1 = \frac{M_{\varphi_1}}{L_{\varphi_1}}\, d\theta + \left(M_\varphi - \frac{M_{\varphi_1} L_\varphi}{L_{\varphi_1}}\right) d\varphi,$$

$$d\varphi_1 = \frac{1}{L_{\varphi_1}}\, d\theta - \frac{L_\varphi}{L_{\varphi_1}}\, d\varphi.$$

This gives

$$J = -\frac{M_\varphi}{L_{\varphi_1}} = -\frac{[F(\varphi_1) - F(\varphi)]\, G'(\varphi) - [G(\varphi_1) - G(\varphi)]\, F'(\varphi)}{[F(\varphi_1) - F(\varphi)]\, G'(\varphi_1) - [G(\varphi_1) - G(\varphi)]\, F'(\varphi_1)}.$$

But $F(\varphi_1) - F(\varphi)$, $G(\varphi_1) - G(\varphi)$ are proportional to $\cos\alpha$, $\sin\alpha$ respectively, while also

$$F'(\varphi) = \cos\tau, \quad G'(\varphi) = \sin\tau, \quad F'(\varphi_1) = \cos\tau_1, \quad G'(\varphi_1) = \sin\tau_1,$$

so that finally we obtain

$$J = \frac{\sin(\alpha - \tau)}{\sin(\tau_1 - \alpha)} = \frac{\sin\theta}{\sin\theta_1}$$

as was stated.

9. Applications to billiard ball problem. As has been seen, there are no points of the ring which are invariant under T. On the other hand consider T^2 followed by a rotation of the θ, φ plane through an angle -2π, which we designate by R_{-1}. The resultant transformation of the ring admits the same area integral as T of course, but advances the points of the outer circle by an angle 2π, and those of the inner circle by an angle -2π of opposite sign. These are the two conditions essential for the application of Poincaré's geometric theorem. Hence $T^2 R_{-1}$ (the compound transformation) possesses two invariant points. This means that T^2 has two geometrically distinct invariant points of oppositely signed indices,* although these correspond to an increase of 2π for φ.

* See my paper *An Extension of Poincaré's Last Geometric Theorem*, Acta Mathematica, vol. 47 (1926). By the index of an invariant point is meant the number of positive rotations of a line joining a point P to its image P_1, when P makes a small positive circuit of the invariant point.

If P is such an invariant point, so is $T(P)$ of course but with the same index. Thus we get two pairs of points, say

$$P,\ T(P),\ Q,\ T(Q),$$

all four distinct. These evidently correspond to the two fundamental periodic motions.

For the application of the theorem of Poincaré to the periodic motions of more complicated type, it is necessary to take account of the fact that every such motion is associated with a distinct second such motion, obtained by reversing the direction of description, although these motions have the same index. However one of these motions increases φ by $2k\pi$, the other increases it by $2(n-k)\pi$. By only considering invariant points of T^n for which φ increases by $2k\pi$, $(k \leqq n/2)$, we clearly obtain each harmonic n-sided polygon only once. It may be noted in passing that this pairing of motions in the billiard ball problem is fully reflected in the fact that T is a product of two involutory transformations; it was the same special property of the ring transformation in the restricted problem of three bodies which enabled me to prove the existence of infinitely many symmetric periodic orbits.*

Now turn to the invariant points of the compound transformation $T^n R_{-k}$ where R_{-k} denotes k-fold rotation through the angle -2π. The rotations on the outer and inner circles are clearly

$$2(n-k)\pi \quad \text{and} \quad -2k\pi,$$

which will be of opposite sign. Thus we can infer the existence of at least two geometrically distinct series of points

$$P,\ T(P), \cdots, T^{n-1}(P), \qquad Q,\ T(Q), \cdots, T^{n-1}(Q)$$

such that $T^n(P) = P$, $T^n(Q) = Q$, while φ has been increased by $2k\pi$; it is assumed that k and n are relatively prime.

* See my paper *The Restricted Problem of Three Bodies,* Rendiconti di Palermo, vol. 39 (1915), in particular, section 14.

To prove this assertion in detail, we may let P be one such invariant point, such that T^n increases φ by $2k\pi$. If the n points

$$P,\ T(P),\ \cdots,\ T^{n-1}(P)$$

are not distinct, let $T^m(P) = P$, $(m \leqq n-1)$, and suppose that φ is increased by $2j\pi$. By combination of the two symbolic equations $T^m(P) = P$, $T^n(P) = P$, we obtain $T^d(P) = P$ where d ($\neq 1$ of course) is the greatest common divisor of m and n. Thus P is invariant under T^d. Suppose that under T^d, the φ of P increases by $2f\pi$. From the equation $T^n = T^{qd}$, we see that T^n will then increase the φ of P by $2qf\pi$, so that $k = qf$. Thus k and n would possess a common factor q, contrary to hypothesis.

Not only are the n points distinct but they have the same index. Hence there will be a point φ invariant under T^n and with oppositely signed index. This, with its images under successive powers of T, will necessarily be distinct from the points of the set generated by P, and leads to a second distinct series of n points.

Hence we obtain for every $n > 2$ and every relatively prime $k \leqq n/2$, two geometrically distinct, harmonic polygons with n sides and making k circuits of the curve C. Corresponding to these there will be of course four periodic motions. We shall not attempt to develop here the characteristics as to type of stability and instability, dependent upon the sign of the index.

It is worthy of observation that the method of sections 2, 3 is evidently applicable here to show that there exist infinitely many periodic motions lying in the vicinity of any periodic motion of general stable type if the constant l is not 0.

We shall indicate in particular how this same method seems to apply to the limiting type of periodic motion in which the billiard ball is rolled around the table.

For this purpose it is essential to examine the explicit formulas given for T in the case when θ is small. A direct computation leads to the result

$$\theta_1 - \theta = \frac{2}{3}\frac{k'}{k^2}\theta^2 + l\theta^3 + \cdots,$$

$$\varphi_1 - \varphi = \frac{2}{k}\theta - \frac{4k'}{3k^3}\theta^2 + m\theta^3 + \cdots,$$

where the function $k(\varphi)$ denotes the curvature of C at the point with given φ, and where the functions $l, m, \cdots$ depend on φ only. Proceeding entirely formally and replacing $\theta_1 - \theta$ and $\varphi_1 - \varphi$ by $\Delta\theta$ and $\Delta\varphi$ respectively, we obtain the approximate differential equation

$$\frac{d\theta}{d\varphi} = \frac{1}{3}\frac{k'}{k}\theta,$$

which gives by integration

$$\theta = \theta_0 k^{1/3}(\varphi).$$

Here θ_0 is the value of θ for a point of curvature unity. This result indicates that, to a first approximation, the curve $\theta = \theta_0 k^{1/3}(\varphi)$ near the inner boundary $\theta = 0$ of the ring is nearly invariant under T, and can probably be modified by the inclusion of higher order terms so as to be still more nearly invariant. Evidently the limiting periodic motions formed by C are to be regarded as analogous to stable periodic motions on this account.

Also if the variable n represents the number of iterations, we have the appreximate differential equation

$$\frac{d\varphi}{dn} = 2\theta_0 k^{-2/3},$$

whence by integration

$$n = \int k^{2/3} d\varphi / 2\theta_0.$$

It follows that φ will increase by more than 2π along the approximately invariant curve if $n\theta_0$ exceeds $\pi K^{2/3}$ where K denotes the maximum curvature of C.

It thus appears as highly probable that the lemma of section 2 is applicable and that there exist infinitely many periodic motions uniformly near to C.

10. The geodesic problem. Construction of a transformation TT^*. We turn our attention next to the problem afforded by the geodesics on an analytic surface. In order to obtain as concrete results as possible, we shall restrict the surface S to be closed and convex, although it is evident that these limitations are not altogether essential for the argument which follows.

We have already established the existence of at least one closed geodesic g of minimax type (chapter V, section 6), which we shall assume to be without double points. Our first assertion is that there exists a positive quantity L so great that any geodesic arc of length exceeding L intersects g at least once (or falls along g). In the contrary case there would exist a sequence of geodesic arcs g_n of length L_n with $\lim L_n = +\infty$, such that each arc g_n does not meet g. This fact would require that for L_n large enough no part of g_n is nearly coincident with g, since nearby geodesics meet a closed geodesic of minimax type in a succession of points separated by arcs of limited length; more precisely, it may be proved that if P is any point of g and P'' is its second conjugate point in the sense of the calculus of variations, then the arc PP'' constitutes at least one complete circuit of g.†

Hence if we let P_n be the mid-point of q_n, the sequence of points P_n will have a limiting point P not on g, and the geodesic q_n will have at least one limiting direction at P, such that the complete geodesic h through P in this direction fails to meet g and indeed nowhere approaches g.

Consider now that part of the surface S (which is divided by g in two parts) upon which h lies, and in particular the part of S lying between g and h. One boundary of this region s is g, of geodesic curvature 0 everywhere, while the other boundary γ consists of part or all of h, and of its limit points.

Let N_1 and N_2 be two nearby accessible points of the

† For a proof see my paper, *Dynamical Systems with Two Degrees of Freedom*, Trans. Amer. Math. Soc., vol. 18 (1917), in particular section 19.

boundary γ so obtained, so that a short curvilinear arc $N_1 A N_2$ exists in the region s with all of its interior points not on the boundary. Consider also the short geodesic arc $N_1 G N_2$. Then $N_1 A N_2 G N_1$ delimits a region σ.

If any interior point of $N_1 G N_2$ is part of the boundary γ, all of $N_1 G N_2$ must be part of it; otherwise the curve h would cut across $N_1 G N_2$ and so certainly intersect $N_1 A N_2$, contrary to hypothesis. In this case $N_1 G N_2$ is a part of the boundary γ and the region σ lies entirely in s. If there is no such interior point, the geodesic arc $N_1 G N_2$ lies inside of s except at N_1 and N_2. Hence if we surround the boundary γ by a cyclical chain of nearby points

$$N_1, N_2, \cdots, N_n,$$

the short distinct geodesic arcs

$$N_1 N_2, N_2 N_3, \cdots, N_n N_1$$

lie inside of s or coincide with γ. We assume that this chain encircles γ in a positive angular sense: in this event, the part toward γ will lie everywhere to the left.

Now in the geodesic polygon so formed, the angle in s at any vertex will be less than or equal to π. For if the angle at N_{i+1}, say, exceeds π and we consider the short geodesic arc $N_i N_{i+2}$, it is apparent that γ lies to the left of $N_i N_{i+2}$ also, so that N_{i+1} is completely encircled by points not on γ which is impossible.

But the integral curvature of the part of S bounded by this polygon will be precisely the sum of these n interior angles diminished by $(n-2)\pi$ by a well known formula, and so will be less than 2π, which is impossible since the integral curvature of either part of S bounded by g is exactly 2π by the same formula.

Thus a contradiction has been obtained, so that the original assertion must be true.

We now introduce parameters as follows. Let an arbitrary directed geodesic f cut the directed geodesic g in a point P.

The position of P may be measured by means of the arc length θ from a fixed point P_0 of g; if the total length of g be taken as 2π by an appropriate choice of a unit of length, the variable θ is a periodic variable of period 2π. Furthermore let φ denote the angle between the positive directions of g and f at the point P so that $0 < \varphi < \pi$.

The next crossing of g by h will be in the opposite sense, and the corresponding θ_1^* and φ_1^* will vary analytically with θ and φ. Thus a transformation T is defined

$$T: \quad \varphi_1^* = f(\varphi, \theta), \qquad \theta_1^* = g(\varphi, \theta),$$

which takes the ring

$$R: \quad 0 < \varphi < \pi, \qquad 0 \leqq \theta < 2\pi$$

(φ, θ, polar coördinates) into itself.

Similarly a crossing in the opposite sense at φ^*, θ^*, will be followed by a crossing φ_1, θ_1, thus defining a second one-to-one, analytic transformation

$$T^*: \quad \varphi_1 = f^*(\varphi^*, \theta^*), \qquad \theta_1 = g^*(\varphi^*, \theta^*)$$

of R into itself.

Along and near the boundaries of R, T and T^* are to be regarded as continuous. This fact may be seen as follows: If a geodesic f near to g intersects g at a small angle φ with given θ, then of course f will intersect later at a small angle φ_1^* with coördinate θ_1^* nearly that of the conjugate point to the first point of intersection, as was observed above. Obviously this ensures the specified continuity. Furthermore, we know that three successive conjugate points correspond to an arc of more than one complete cycle of g.

If we have $\varphi = 0$ or π we have respectively $\varphi_1^* = 0$ or π, while if $\varphi^* = 0$ or π we have likewise $\varphi_1 = 0$ or π. Furthermore as θ or θ^* increases along a boundary so will θ_1^* or θ_1; in fact it has been noted that if θ is the coördinate of P, then θ_1^* is the coördinate of the conjugate point P'. Hence

T and T^* are direct transformations leaving the outer and inner boundaries of R invariant.

It is now desirable to consider somewhat more in detail the nature of T and T^* along the boundaries. These transformations are evidently not determined up to a complete rotation, and it is desirable to make a convention which eliminates this arbitrary factor, and enables us to compare the transformations along the inner and outer boundaries. Let us consider any directed geodesic arc $P_0 P_1$ of f between two successive crossings P_0 and P_1 of g by f, and let us define a double point Q of $P_0 P_1$ as positive if a moving point in describing the arc $P_0 P_1$ passes over the arc $P_0 Q$ already described at Q, from the left to the right side; and as negative in the contrary case. The 'index' of $P_0 P_1$ may be defined as the difference between the number of positive and the number of negative crossings within $P_0 P_1$. Let us define the value θ_1^* of θ at P_1 as that corresponding to the positively taken arc $P_0 P_1$ increased by $2\pi i$ where i is the index.

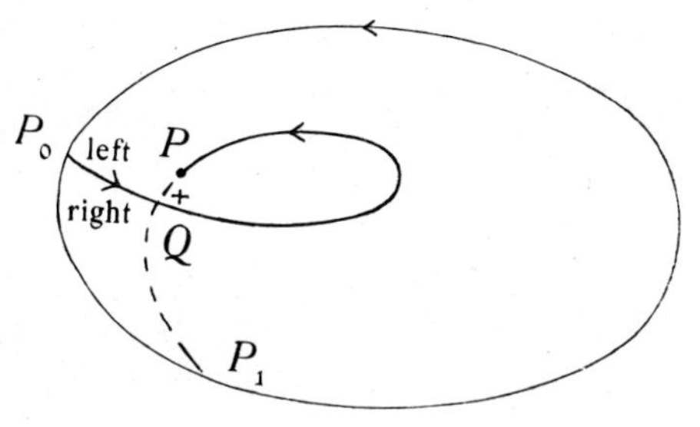

We can now show that, as thus defined, θ_1^* varies continuously with θ. In fact as θ and φ change, no new crossings can be introduced within $P_0 P_1$ inasmuch as a geodesic arc cannot be tangent to itself. If a positive crossing is introduced at P_1, evidently θ_1^* increases through an exact multiple of 2π as it should. If a negative crossing is introduced, θ_1^* decreases similarly. Thus the convention is appropriate to all cases.

Now it is also evident that if a geodesic arc $P_0 P_1$ near to g be continuously deformed near g while not maintaining its geodesic character, with P_0 and P_1 fixed, and only simple interior tangency allowed, then positive and negative crossings appear in or disappear in associated pairs. Thus if the arc $P_0 P_1$ be deformed into a spiral so as to eliminate all negative crossings, it is plain that the index will equal the number of apparent circuits of g made by the spiral. In this case the increase in θ, namely $\theta_1^* - \theta$, will be measured by the arc length along the curve g, according to our convention. However if the positive crossings are eliminated, the convention gives $2\pi i$ $(i \leqq 0)$ increased by the angular value of the short positive arc $P_0 P_1$ as the difference $\theta_1^* - \theta$, and hence this difference exceeds the arc length $P_0 P_1$ by 2π.

Consequently it is seen that $\theta_1^* - \theta$ is measured in the sense of the convention by the actual increase in θ along g for $\varphi = 0$, while, for $\varphi = \pi$, $\theta_1^* - \theta$ is measured by the algebraic increase along the arc (actually a decrease) augmented by 2π since $P_0 P_1$ is to be taken positively.

Now in going from a point P to its second conjugate point P'' with $\varphi = 0$, θ increases by an amount α,

$$2k\pi \leqq \alpha < 2(k+1)\pi,$$

where $k \geqq 1$ is independent of the position of P; in fact the one-to-one, direct transformation of the points of g from P to P'' defines a rotation number† which lies between $2k\pi$ and $2(k+1)\pi$ with $k \geqq 1$, because of a property of the conjugate points along a geodesic of minimax type already specified. Hence we infer

$$2k\pi \leqq g(0, \theta) - \theta < 2(k+1)\pi.$$

Similarly if we consider $\varphi = \pi$, then $\theta_1^* - 2\pi$ is diminished by a like amount so that we find

$$-2k\pi < g(\pi, \theta) - \theta \leqq -2(k-1)\pi.$$

It follows that the transformation T (or T^*) advances points in opposite angular directions along the two boundaries of R,

† For definition and short discussion of the Poincaré rotation numbers, see my paper, *Surface Transformations and their Dynamical Applications*, Acta Mathematica, vol. 43 (1922), section 45.

at least if the exceptional case of a second conjugate point coincident with the initial point after a single circuit be excluded.

11. Application of Poincaré's theorem to problem. The geodesic problem is of Hamiltonian form of course, with principal function H given by the squared velocity. In the four-dimensional manifold of states of motion, the quadruple integral

$$\int\int\int\int dp_1\, dq_1\, dp_2\, dq_2$$

is an invariant integral. In a particular invariant sub-manifold $H =$ const. there must then be an invariant volume integral, namely

$$\int\int\int dq_1\, dp_2\, dq_2,$$

provided that $H = h, q_1, p_2, q_2$ are taken as coördinates (section 3). The restriction $H =$ const. merely fixes the constant velocity.

The ring R of states of motion crossing g positively is evidently represented in this manifold by a ring $\bar{R}$ bounded by the two closed curves representing g traversed in the two possible senses. The transformed point $TT^*(P)$ of a point P of this ring by TT^* is obviously obtained by following the corresponding stream line until it meets $\bar{R}$ a first time. Further consideration shows that the ring $\bar{R}$ so obtained is an analytic surface.

Now if we consider a tube of stream lines with two bases of areas $\Delta\sigma_1$, $\Delta\sigma_2$, in $\bar{R}$, with q_1 as independent variable, and if α_1, α_2 be the angles which the stream lines make with $\bar{R}$ at these two surface elements respectively, then the loss of 'volume' at one base in time Δq_1 is nearly

$$(\sin\alpha_1\, \Delta\sigma_1)\, \Delta q_1$$

while the gain at the other is nearly

$$(\sin\alpha_2\, \Delta\sigma_2)\, \Delta q_1 .$$

Thus $\int \sin \alpha \, d\sigma$ yields the positive invariant area integral necessary for the application of Poincaré's geometric theorem.

Hence there exist two points of R invariant under TT^*, and this gives immediately the following conclusion:

Let there be given a convex analytic surface on which the closed geodesic of minimax type known to exist is without double points, and suppose that the second conjugate point of no point on this geodesic arises on precisely a single complete circuit. Then there will exist a second closed geodesic which intersects the known minimax geodesic only twice.

Of course a single closed geodesic yields two invariant points.

Under the same conditions there must be two distinct closed geodesics which meet the geodesic of minimax type only twice.

To prove this we proceed as follows. We have for the points of R

$$T(\theta, \varphi) = (\theta_1^*, \varphi_1^*)$$

by definition. On the other hand the same geodesic may be taken in the opposite sense so that

$$T(\theta_1^*, \pi - \varphi_1^*) = (\theta, \pi - \varphi).$$

If then we define the 'reflection' U in such wise that

$$U(\theta, \varphi) = (\theta, \pi - \varphi),$$

we obtain

$$TU(\theta, \varphi) = (\theta_1^*, \pi - \varphi_1^*)$$

and thence

$$TUTU = I$$

where I is the identity. Hence $TU = V$ is a transformation of period 2, as is U, and we find $T = VU$, i. e. T is a product of two involutory transformations. Similarly we find $T^* = V^*U$ where V^* is also involutory. Hence we infer that TT^* has the form VUV^*U. Suppose now that there is an invariant point P under TT^* so that

$$VUV^*U(P) = P.$$

We obtain then, by inverting,

$$UV^* UV(P) = P,$$

whence

$$TT^* V(P) = V(P).$$

Hence if P is an invariant point under TT^*, so is $V(P)$.

But $V(P)$ must be a geometrically distinct point from P itself. Otherwise we should have

$$TU(P) = P$$

or, more explicitly, for the corresponding (θ, φ)

$$\theta_1^* = \theta, \qquad \pi - \varphi_1^* = \varphi.$$

But this would mean that the first intersection of the geodesic with the minimax geodesic g crosses it at the same point and in the opposite direction, which is manifestly impossible.

Now the indices of the invariant point P and of the associated invariant point $V(P)$ under the transformation TT^* are equal. In fact make the change of variables corresponding to the symbolic equation

$$Q = V(P),$$

by which any point P is taken into $V(P)$. The modified transformation less is of course

$$VTT^* V = (TT^*)^{-1}$$

as is at once verified by substitution of the factored form of TT^* derived earlier. Hence the transformation TT^* in the neighborhood of one invariant point is equivalent to the inverse transformation in the neighborhood of the associated invariant point. But the index of an invariant point is unaltered by a change of variables, and is the same as for the inverse transformation. Consequently the indices of the two associated invariant points are necessarily equal in all cases.

It is obvious geometrically that the association of pairs of points arises from the circumstance that every geodesic may be traversed in two opposite senses.

Since the extended theorem of Poincaré allows us to infer the existence of two invariant points with oppositely signed indices, we infer that there exist two geometrically distinct closed geodesics which intersect the closed geodesic only twice.

This completes the proof of the previous italicized statement.

By applying the same theorem to higher powers of TT^* the existence of other types of closed geodesics could be inferred. Moreover the methods of section 1 are applicable and show that there will be in general infinitely many closed geodesics in the immediate vicinity of a closed geodesic of stable type.

There are two further remarks concerning the geodesic problem which I will make in conclusion. In the first place the conclusion that there will exist at least two other distinct closed geodesics meeting the geodesic of minimax type only twice is no doubt valid for surfaces of much more general type.

Secondly, if the convex surface is symmetric in space about a plane containing a geodesic g, there will be various closed geodesics which intersect g twice at right angles. Methods for dealing with these symmetric closed geodesics can be used analogous to those which I employed in dealing with certain symmetric periodic orbits in the restricted problem of three bodies (loc. cit.). In fact if g is of minimax type, the transformations T and T^* become identical, and TT^* appears as the square of a product of two involutory transformations, as is the case for the fundamental transformation in the restricted problem of three bodies.

CHAPTER VII

GENERAL THEORY OF DYNAMICAL SYSTEMS

1. Introductory remarks.* The final aim of the theory of the motions of a dynamical system must be directed toward the qualitative determination of all possible types of motions and of the interrelation of these motions.

The present chapter represents an attempt to formulate a theory of this kind.

As has been seen in the preceding chapters, for a very general class of dynamical systems the totality of states of motion may be set into one-to-one correspondence with the points, P, of a closed n-dimensional manifold, M, in such wise that for suitable coördinates $x_1, \cdots, x_n$, the differential equations of motion may be written

$$dx_i/dt = X_i\,(x_1, \cdots, x_n) \qquad (i = 1, \cdots, n)$$

in the vicinity of any point of M, where the X_i are n real analytic functions and where t denotes the time. The motions are then presented as curves lying in M. One and only one such curve of motion passes through each point P_0 of M, and the position of a point P on this curve varies analytically with the variation of P_0 and the interval of time to pass from P_0 to P. As t changes, each point of M moves along its curve of motion and there arises a steady fluid motion of M into itself.

By thus eliminating singularities and the infinite region, it is evident that we are directing attention to a restricted class of dynamical problems, namely those of 'non-singular' type.

* Sections 1–4 are taken directly from my paper *Über gewisse zentrale Bewegungen dynamischer Systeme*, Göttinger Nachrichten (1926). The remainder is closely related to my papers, *Quelques théorèmes sur les mouvements des systémes dynamiques*, Bull. Soc. Math. France, vol. 40 (1912), *Surface Transformations and Their Dynamical Applications*, Acta Mathematica, vol. 43 (1912), in particular sections 54–57.

However, most of the theorems for this class of problems admit of easy generalization to the singular case. The problem of three bodies, treated in chapter IX, is of singular type.

The differential equations of classical dynamics are more special, and in particular possess an invariant n-dimensional integral over M. In consequence any small molecule, σ, of M about a point P_0 at time t_0, must subsequently overlap its first position σ_0. This fact may be deduced as follows: Suppose that at a time t_1 following τ units after t_0, the molecule is in an entirely distinct position, and consider its positions at times $\tau, 2\tau, \cdots$ after the initial instant t_0. These positions cannot be entirely distinct from one another; for if v denotes the value of the invariant integral over the molecule in its initial position, this value will be the same in such subsequent positions, and since the value of the invariant integral over M is finite, say V, the number of distinct positions cannot exceed V/v. Hence some i-th and j-th molecules overlap $(i<j)$. But if these overlap, then in the corresponding positions $(j-i)\tau$ units of time earlier, they will still do so. It follows that the $(j-i)$-th position of σ overlaps σ_0. By this argument and its natural extension, Poincaré* proved that in general the motions of such more special dynamical systems will recur infinitely often to the neighborhood of an initial state, and so will possess a kind of stability 'in the sense of Poisson'.

It will be our first aim in this chapter to show that with an arbitrary dynamical system not so restricted, there is associated always a closed set of 'central motions' which do possess this property of regional recurrence, towards which all other motions of the system in general tend asymptotically.

2. Wandering and non-wandering motions. Consider an arbitrary point P_0 of the manifold M of states of motion. Let σ be an open continuum of small diameter† ε, containing

* *Méthodes nouvelles de la Mécanique céleste,* vol. 3, chap. 26.

† It is evident that distance may be defined in an appropriate fashion in M. The diameter of a set of points is merely the upper limit of distances between pairs of points of the set.

P_0. As the time t increases, this 'molecule' σ moves. It may happen that P_0 represents a state of equilibrium; in that case the molecule will always continue to overlap P_0; we exclude this case for the moment. In any other case σ will move outside of itself if σ_0 is small enough, since the velocity components dx_i/dt are approximately the same as at P_0 throughout the molecule. If it is possible to take ε so small that σ never again overlaps its first position, we shall call P_0 a 'wandering point', and the corresponding motion a 'wandering motion'. In the contrary case P_0 will be termed a non-wandering point, and the corresponding motion a non-wandering motion. With this second class we naturally include equilibrium points and the degenerate corresponding motions.

There is an apparent asymmetry between the increasing and decreasing directions of the time t, as far as these definitions go. But it is easily seen that there is no asymmetry in actuality. In fact if the image of σ intersects itself after τ units of time, it does τ units of time earlier; for the overlapping molecules σ and σ_τ (i. e. taken τ units afterward) occupy the positions $\sigma_{-\tau}$ and σ respectively, τ units of time earlier, and these continue to overlap.

Thus the wandering point P_0 is characterized by the fact that the corresponding molecule σ describes an n-dimensional tube which never overlaps itself as t changes from $-\infty$ to $+\infty$. For this reason the characterization as 'wandering' seems legitimate, since P_0 never recurs to the infinitesimal neighborhood of any points once passed.

The set W of wandering points of M is made up of curves of motion filling open n-dimensional continua. The set M_1 of non-wandering points of M is made up of the complementary closed set of curves of motion.

For the reasons just presented all of this statement is obvious, except perhaps for the assertion that W is open and consequently M_1 is closed. But if P_0 is a wandering point, so evidently are all the points of the molecule σ including P_0. This shows at once that W is made up of open continua, and hence that M_1 is closed.

If the set M_1 of non-wandering motions contains points which are not limit points of the set W, these form a sub-set M_1' of curves of motion filling up open n-dimensional continua and possessing the property of regional recurrence.

It is clear that M_1' is made up of a set of curves of motion, for since Q of M_1' is not in the immediate neighborhood of any curve of motion of W the same will be true of all other points on the curve of motion containing Q. Also a sufficiently small molecule containing Q will be entirely in M_1' so that M_1' is made up of open n-dimensional continua of non-wandering points. Hence the property of regional recurrence is obvious.

Evidently the set $M_1 - M_1' = M_1''$ is merely the set of boundary points of the n-dimensional open continua W, M_1'. It is made up of a closed set of complete motions, of less than n dimensions.

As time increases or decreases, every wandering point approaches the set M_1 of non-wandering curves of motion.

The proof of this fundamental property of wandering motions is not difficult.

Consider any small open neighborhood of M_1, within which M_1 lies, and the complementary closed set C made up exclusively of the points W. About each point of C can be constructed a small molecule σ, such that the molecule never overlaps itself as time changes. Hence a finite set of these molecules can be found which cover C completely. A moving point can enter any one of these molecules (held fixed) only once, and can stay within it only a short interval of time. It is thus obvious that after a finite time the moving point lies always inside of the arbitrary neighborhood of M_1. Hence any moving point must approach M_1, as stated.

A more detailed study reveals certain further characteristics of the mode of approach of the wandering motions to the non-wandering motions. Since in the above discussion the moving point enters one of the fixed molecules covering C only once and stays in it for only a brief period of time, the following facts are obvious.

*Any wandering motion remains outside of a prescribed neighborhood of M_1 only a finite time T, and goes out of this neighborhood only a finite number of times N, where N and T are uniformly limited, once the neighborhood is chosen.**

3. The sequence $M, M_1, M_2, \cdots$. Having arrived at the closed set of non-wandering points M_1, between points of which distance may be defined as in M, we are in a position to define wandering and non-wandering points relative to M_1, (instead of M) as follows. Choose an arbitrary point P_0 of M_1, and an open continuum σ, of small diameter, containing P_0 and certain other points of M_1. Setting aside the case when P_0 is an equilibrium point, and choosing the diameter of σ small enough, we see that this molecular part of M_1 will move outside of itself away from its initial position. If the diameter can be chosen so small that the part of M_1 in σ never again overlaps itself, we may say that P_0 is a wandering point of M_1 (though of course non-wandering with respect to M by definition of M_1). Other points, including equilibrium points, in M_1 may be termed non-wandering points of M_1.

It is clear that the parallelism is complete. The non-wandering points M_2 relative to M_1 form a closed set of motions toward which any point P of the set of wandering points W_1 is asymptotic as time increases or decreases, and similar uniformity properties relative to M_1 hold.

The same process may now be repeated with respect to M_2 as a basis, and thus are defined M_3 and W_2. Continuing in this manner we obtain $M, M_1, M_2, \cdots$. We regard the process as terminating if any M_{i+1} is the same as M_i, in which case there are no points W_i of course. In case the process does not terminate in this way the sequence of distinct closed sets $M, M_1, M_2, \cdots$, each within its predecessors, defines a unique limiting set M_ω, which is evidently closed

* To render the count of exits precise, it is desirable to take each covering molecule so as to have at most one segment cut off by any neighboring stream line, and then to consider the neighborhood outside of the covering molecules as the given neighborhood of M_1.

and composed of curves of motion. Further application of the process yields $M_{\omega+1}, M_{\omega+2}, \cdots$. Thus are defined successively

$$\begin{array}{llll} M, & M_1, & M_2, & \cdots, \\ M_\omega, & M_{\omega+1}, & M_{\omega+2}, & \cdots, \\ M_{2\omega}, & M_{2\omega+1}, & M_{2\omega+2}, & \cdots, \\ \cdot\ \cdot\ \cdot & \cdot\ \cdot\ \cdot & \cdot\ \cdot\ \cdot & \cdot\ \cdot\ \cdot \\ M_{\omega^2}, & M_{\omega^2+1}, & M_{\omega^2+2}, & \cdots, \\ \cdot\ \cdot\ \cdot & \cdot\ \cdot\ \cdot & \cdot\ \cdot\ \cdot & \cdot\ \cdot\ \cdot \end{array}$$

in accordance with the well known theory of transfinite ordinals given by Cantor.

But these form an ordered set of distinct closed point sets each with an immediate successor and contained in all of its predecessors. Such a set is certainly numerable. Hence the process does eventually terminate in some M_r.

Thus there exists a well-ordered, terminating set of distinct closed sets

$$M,\ M_1,\ M_2,\ \cdots,\ M_\omega,\ \cdots,\ M_r,$$

in which an element M_{p+1} immediately following M_p consists of the non-wandering motions relative to M_p, while an element M_p without an immediate predecessor is the limit of its predecessors. The wandering points W_p of M_p tend asymptotically toward the non-wandering motions M_{p+1} in such wise that the total time outside of a given neighborhood of M_{p+1}, as well as the number of exits from this neighborhood, are uniformly limited.

The final set obtained, M_r, is the set of 'central motions'. It is obvious that these have the property of regional recurrence since there are no wandering points W_r. From this property it may be inferred by the method of Poincaré (loc. cit.) that in any arbitrary neighborhood of a point of M_r there is a motion which enters this region infinitely often in past and future.

In fact, if there is an isolated curve of motion of M_r in the region, the curve must be closed and this motion must be itself periodic, since the motion is non-wandering. In this case the periodic motion itself has the desired property. If there is no isolated motion, we can select a neighborhood of the given point as small as we please which overlaps it in M_r again at least once at some later time. Thus we get two points P, Q, on the same curve of motion of M_r, both in the molecule about the given point, but not near together in time. Next we choose a still smaller molecule about P, so small that every point P' in this molecule continues to meet the original molecule at a point Q' near to Q. But by choosing P' suitably, we can find a point R' before P' in time, lying in the same molecule as P', Q'. Thus we obtain an arc $R' P' Q'$ of a curve of motion of M_r such that the three points R', P', Q' lie in the given molecule taken at the outset at three different times. Next by choosing a still smaller molecule about P' we are led to an arc $R'' P'' Q'' S''$, then to $T^{(3)} R^{(3)} P^{(3)} Q^{(3)} S^{(3)}$, and so on. The limit of the points $P, P', P'', \cdots$, will be a point of M_r in the given molecule, which traverses that molecule infinitely often in the past and in the future.

It is obvious that the periodic motions in the dynamical problem must lie in the set of central motions. The motions defined later (section 7) as 'recurrent' will also.

4. Some properties of the central motions. It is easy to see that every point of M approaches within an arbitrarily given neighborhood of the central motions at least once within every sufficiently large fixed interval of time. For such a point certainly approaches M_2 in this manner inasmuch as every motion of M_1 approaches M_2 uniformly often, and therefore every motion near M_1 does also. Continuing in this manner we see that this uniformity property holds for $M_1, M_2, \cdots$. If the series continues to M_ω, the same property must be true for M_ω. Indeed an M_n can then be found in any neighborhood of M_ω since M_ω is the inner limiting set of the closed sets M_n. Since a point approaches an arbitrarily given neighborhood of M_n uniformly often, it is therefore evident that it does the same for M_ω. By continuing indefinitely in this manner with $M_{\omega+1}, \cdots, M_{\omega^2}, \cdots, M_r$ we arrive at the stated conclusion.

Now the motions of M were seen to approach those of M_1 in a definite manner, while in turn those of M_1 approach M_2 similarly. By combining these results it would be possible to make certain statements as to the mode of approach of any point of M to M_2. This method of description might be carried on to $M_3, \cdots, M_\omega, \cdots$. But for the general purpose of this paper we shall merely establish a simpler result for all the sets M_p.

Let us define the 'probability' that an arc PQ lies in a given region Σ as the ratio of the time interval for the part that does lie in Σ to the total interval.

The probability that an arc of a curve of motion lies inside of an arbitrary neighborhood of a set M_p, and in particular the set of central motions M_r, approaches unity uniformly as the interval of time for such an arc increases indefinitely.

From what was proved at the outset concerning M_1, the probability that any arc lies in a given neighborhood of M_1 approaches unity uniformly as the interval of time increases indefinitely.

To prove a like result for M_2, we recall that any arbitrary motion of M_1 is represented by a curve which lies inside of a prescribed neighborhood of M_2 except for a limited number of arcs corresponding to a limited total interval. Now any long arc sufficiently near to M_1 will share the property of the motion of M_1, of being inside the prescribed neighborhood except for a limited number of arcs of limited total length, provided that first the length of the long arc is taken arbitrarily large, and then the neighborhood of M_1 within which it is to lie is chosen suitably. Furthermore if a point is near enough to M_1 it will evidently lie on such a long arc of a curve of motion.

Hence, since the probability that an arc of M lies within the prescribed neighborhood of M_1 approaches unity as the time interval is taken longer and longer, and since every point in such a neighborhood of M_1 is part of a long arc near to M_2 except for a finite number of pieces of limited total length, it is obvious that the probability that an arc in M lies within a given neighborhood of M_2 approaches unity uniformly as the time interval for the arc becomes larger and larger.

This same argument is evidently applicable for $M_3, M_4, \cdots$. For M_ω we need only note that since M_ω is the inner limiting set defined by $M_1, M_2, \cdots$, it will be approached as a limit by this set in such a way that for a sufficiently large n every point of M_n will be within distance ε of some point of M_ω. Since the probability that an arc of an arbitrary motion for a sufficiently long interval is within distance ε of M_n is at least $1 - \delta$ where δ is small, the probability that it is within distance 2ε of M_ω is at least $1 - \delta$.

Evidently this reasoning admits of indefinite continuation and leads to the desired conclusion.

5. Concerning the role of the central motions. It is obvious then that a first problem concerning the properties of dynamical systems is the determination of the central motions.

For the equations of classical dynamics the central motions are obviously the totality of motions, at least for the case without singularities to which we are now confining our attention. In fact the property of stability in the sense of Poisson involves that of regional recurrence, characteristic of the central motions.

The superior usefulness of the equations of classical type may very well be a reflection of the fact that the central motions are the most probable motions, rather than any consequence of the laws of nature.

6. Groups of motions. Consider now any curve of motion in M with a point P_t moving on it. The points P_t constitute the 'point group' of the given motion.

Every limit point of the set P_t for $\lim t = +\infty$ will be termed an ω limit point of the motion, and every limit point of the set P_t for $\lim t = -\infty$ will be termed an α limit point.

In all cases the limit points of either class form a closed point set.

The set of ω (α) *limit points of any motion* P_t *form a closed connected set of complete motions. The distance of* P_t *from this limit set approaches* 0 *for* $\lim t = +\infty$ $(-\infty)$.

In fact let P^* be an ω limit point which P_t approaches for $\lim t = +\infty$, and let P^{**} be a point of the curve of motion through P^*, after an interval c. Evidently P_{t+c} will approach P^{**} in the same sense. That is to say, P^{**} is an ω limit point if P^* is. From this argument we infer that all points of the point group of P^* are ω limit points.

To establish that the distance from P_t to this set approaches 0, we employ an indirect argument. If P_t did not approach the set of ω limit points uniformly for $\lim t = +\infty$, it would be possible to select an infinite set of indefinitely increasing values of t, such that P_t would be distant from any ω limit point by at least a definite positive quantity d. There would then be at least one limit point P_1 of the set P_t, and this point would be at least d distant from any ω limit point. By definition, however, P_1 is an ω limit point, so that a contradiction results.

It is obvious that the ω limit set is connected inasmuch as it is approached uniformly by the point P_t as t becomes infinite, while P_t moves along its curve of motion continuously.

If we consider the groups of motions $M_1, M_2, \cdots$ which lead to the central motions M_r, it is obvious that the α and ω limit motions of a motion in M_p will form part of M_{p+1}.

7. Recurrent motions. Consider now an arbitrary, closed, connected set Σ of complete motions. It was observed above that the α or ω limit motions of any motion form such a set of motions. More generally, if we take any connected set of complete motions and adjoin to it the limit points, we obtain an enlarged set Σ.

If a set Σ contains no proper sub-set Σ' of the same type we shall say that Σ is a 'minimal set of motions'. In this case if P is any point of Σ, its α and ω limit points form closed sets in Σ, which must therefore coincide with Σ.

By definition any complete point group in a minimal set forms a recurrent point group and any motion in that group is called 'recurrent'.

All recurrent motions belong to the central motions. In fact the α and ω limit points of any such motion in M_p form a set Σ in M_{p+1}, which must coincide with the minimal set, so that no point of the set can be in M_p but not in M_{p+1}. Hence the minimal set corresponding to the recurrent motion lies in M_r.

In all cases but the simplest one, in which Σ consists of a single closed curve, a minimal set Σ contains a non-denumerable perfect set of curves of motion. For suppose a minimal set to have an isolated curve of motion. A point P_t on this curve has the points of the curve as its α or ω limit points. Hence this curve must be closed and constitutes the minimal set Σ.

In order that a point group generated by a motion P_t be recurrent, it is necessary and sufficient that for any positive quantity ε, however small, there exists a positive quantity T so large that any arc $P_t P_{t+T}$ of the curve of motion has points within distance ε of every point of the curve of motion.

This condition is necessary.

If not there is a recurrent point group Σ generated by P_t and a positive ε such that a sequence of arcs $P_t P_{t+2T}$ (T, arbitrarily large) can be found for each of which no point of the arc which comes with distance ε of a corresponding point Q of Σ. As T increases, the point Q has at least one limit point Q^*, and thus it is clear that for a properly taken subset of the sequence $P_t P_{t+2T}$, no point lies within distance $\varepsilon/2$ of Q^*. Consider the sequence of middle points P_{t+T} of such arcs. For a limiting position P^*, we infer that every point of the complete point group of P^* is at distance at least $\varepsilon/2$ from Q^*. Hence P^* defines a closed set of point groups lying within the closed minimal set Σ defining the given recurrent motion, but forming only part of it, and in particular not containing Q^*. This is absurd by the very definition of a minimal set.

To prove the condition sufficient, we note first that the sets of α and ω limit points of a point group satisfying this condition must coincide. We need only take $t = 0$ in the arbitrary arc $P_t P_{t+T}$ to see the truth of this fact. Call the set of these common α and ω limit points, Σ.

If the set Σ is not minimal it would contain a proper subset Σ' of the same sort to which some point Q of Σ would not belong. Now, when the point P_t approaches sufficiently near to a point of Σ', it will remain very near to this closed connected set of complete motions for an arbitrarily long interval of time and so will not approach all of Σ in this interval, as demanded. Thus the assumed condition would not be satisfied by the point group generated by P_t.

Hence Σ is minimal, and the motion is recurrent.

Clearly all recurrent motions are central motions, but of course central motions need not be recurrent. Indeed in the case of the differential equations of classical dynamics, all the motions are central, but need not be recurrent.

8. Arbitrary motions and the recurrent motions. The importance of the motions of recurrent type for the consideration of any arbitrary motion is evidenced by the following result:

There exists at least one recurrent group of motions in the $\omega(\alpha)$ *limit motions of any given motion.*

Let Σ denote the closed set of ω limit points of the given motion. We need to prove that the set Σ contains a minimal sub-set.

Divide M into a large number of small regions of maximum span not greater than ε, an assigned positive constant. Among the motions of Σ there will be one which enters a least number of these small regions of M under indefinite increase and decrease of t. Let Σ_1 be the corresponding closed set of complete limit motions. This set is part of Σ and lies wholly in the same small regions. Divide these small regions into regions of maximum span $\varepsilon/2$. Among the motions of Σ_1 there will be one which enters a least set of these smaller regions of M under indefinite increase and decrease of t.

Define Σ_2 as the corresponding closed set of limit motions, which will be part of Σ_1.

Proceeding in this way we determine an infinite sequence $\Sigma_1, \Sigma_2, \cdots$ of closed connected sets of complete motions, each set being contained in its predecessors. Now let P_n be any point whatever of Σ_n, and let P^* be a limit point of the set P_n. The point P^* belongs to Σ of course since it is a limit point of points of Σ. Furthermore since P_n is contained in Σ_m $(m \leqq n)$, the limit point P^* lies in all of the regions $\Sigma, \Sigma_1, \cdots$. Likewise since the complete curve of motion through P_n is contained in Σ_m $(m \leqq n)$, the complete curve of motion through P^* lies in $\Sigma, \Sigma_1, \cdots$. Hence, in accordance with the property by which these regions were defined, the curve through P^* must enter all of the sub-regions at every stage, and hence its limit points must consist of all the points Σ_r common to $\Sigma, \Sigma_1, \Sigma_2, \cdots$.

The same argument shows that any motion lying in Σ_r has this complete set as its set of α or ω limit points. In other words the set Σ_r forms the desired minimal set.

The following further result shows that either a point P_t generates a recurrent motion, or else that it successively approaches and recedes from such recurrent motions uniformly often:

For any $\varepsilon > 0$ *there exists an interval* T *so large that any arc* $P_t P_{t+T}$ *in* M *contains at least one point within distance* ε *of some group of recurrent motions.*

The proof is immediate.

If the theorem is not true it is possible to obtain arcs $P_t P_{t+2T}$, not coming within distance ε of a recurrent point group for T arbitrarily large. Let then P_{t+T} denote the middle point of such an arc. If P^* is a limit point of the points P_{t+T} for $\lim t = +\infty$, evidently the complete curve through P^* has none of its points within distance ε of any recurrent point group. But the set of α and ω limit points of P^* each contain a minimal set. Thus a contradiction appears, since every motion in a minimal set is by definition recurrent.

9. Density of the special central motions. It is evident that the structure of the set of central motions M_r is of vital theoretic importance. Now this closed set of motions has been seen to be characterized by the property of regional recurrence, and thus the existence of an n-dimensional invariant volume integral for the equations of classical dynamics insures that the totality M_r is M itself in this case.

We propose to establish some simple properties of the set of central motions.

The set M_r is made up of one or more connected parts, each of which contains at least one minimal set of recurrent motions.

Any central motion whose α or ω limit points do not fill up all of the connected part of M_r on which the motion lies will be termed a 'special' central motion. A recurrent motion is special according to this definition unless the corresponding minimal set constitutes all of the connected part of M_r to which the recurrent motion belongs.

In particular then for classical dynamics, the special motions are those which do not pass arbitrarily near all possible states of motion, either as time increases or else as time decreases.

The special central motions are everywhere dense on any connected part of the set M_r of central motions, unless that part is made up of a single minimal set of recurrent motions.

For the case of classical dynamics ($M_r = M$) the special motions are thus dense throughout M, unless M is made up of a single minimal set of recurrent motions.

In establishing this result we shall take M_r as M itself, but it will be evident that the proof applies equally well to any connected part of M_r provided that by a region of M_r is understood any connected part of the set M_r, no point of which is a limit point of points not belonging to the region but in M_r.

Suppose if possible that there is a closed region E no point of which belongs to a special motion. Now there exists at least one set of recurrent motions Σ in M, which are special motions in the case under consideration and so fall entirely within the complementary region $F = M - E$.

Consider all of the points within a distance ε of Σ where ε is chosen so that the ε neighborhood of Σ lies entirely within F.

As t increases indefinitely, this ε neighborhood moves, and two possibilities arise: either (1) no points of the ε neighborhood go outside of F for ε sufficiently small or (2) at least one point of the neighborhood finally emerges from F, no matter how small ε is chosen.

The second alternative is easily disposed of. Let ε approach 0 and consider a sequence of arcs PQ of curves of motion in F such that P lies in the ε neighborhood of Σ, while Q lies on the boundary of F and corresponds to a later time t. Evidently the half curve of motion in the sense of *decreasing* time through any limiting position $\bar{Q}$ of Q for $\lim \varepsilon = 0$, lies wholly in F, and constitutes a special motion of the type whose existence was denied. Thus we may confine our attention to the first alternative.

But in the first case choose ε as large as possible so that the set of motions passing through the ε neighborhood of Σ continue to lie wholly in F as t increases. It is apparent that the upper limit of values ε for which this is true is also a permissible value of ε. The points on these motions and limit motions constitute an augmented region R within which Σ lies. No motion of R when continued in the sense of decreasing time can emerge at a point P on the boundary of F, for then the motion through P for *increasing* time would be a special motion leaving E at P, and thereafter lying within F. For the same reason R must lie wholly within F. But now if we consider the points within the ε' neighborhood of R, some of these must emerge at a later point Q from F; otherwise the region R would not correspond to a maximum value of ε. Thus there arises a limiting point $\bar{Q}$, the motion through which is special of course, and remains in F with decreasing t. Thus a contradiction follows in all cases.

A slight extension of this reasoning enables us to establish the following more precise result:

If a connected region of M_r contains a motion within it, there exists at least one special motion passing through a point of its boundary and lying within the region as t increases (decreases).

Suppose that we consider an ϵ neighborhood of the complete motion within the region F. If t decreases indefinitely, this region moves and our earlier argument shows that the fact stated must be true unless no points of this ϵ neighborhood ever go outside of F with decrease of time, for ϵ small.

If we consider the ϵ neighborhood together with all of the part of F into which it moves as t decreases we obtain an augmented region. This augmented region must lie in F when t increases as well as decreases, and it must be invariant. For if a point of the augmented region moves outside of itself to a point Q as t increases, then a small enough molecule about Q would never overlap Q again as t decreases, contrary to the property of regional recurrence.

If we take ϵ as the upper limiting value, we obtain an invariant region R in F made up of complete motions. By considering points in the ϵ' neighborhood of R and letting t decrease, we find as before that some of the motions in this neighborhood must finally leave F at a point Q. By allowing ϵ' to approach 0, a limit point $\bar{Q}$ is found through which passes a motion lying in F for increasing t, as desired.

10. Recurrent motions and semi-asymptotic central motions. We shall say that a motion is positively (negatively) 'semi-asymptotic' to a minimal set of recurrent motions in case that set is the only minimal set among its ω (α) limit motions. With this definition in mind, we can state the following conclusion:

Either there are other recurrent motions in the immediate vicinity of a recurrent motion, or there exist central motions positively (negatively) semi-asymptotic to the recurrent motion.

The proof is immediate. Choose a small neighborhood of the given minimal set of recurrent motions. By the preceding section there will be a motion entering this neighborhood at a point P, and remaining within it subsequently. If this motion has other minimal sets besides the given minimal set in its set of ω limit points for ε arbitrarily small, the statement made is true. In the contrary case the motion through P will be positively semi-asymptotic to the given recurrent motion, also in accordance with the result stated.

11. Transitivity and intransitivity. Let us consider a 'molecule' about an arbitrary point in some connected part of the manifold of central motions, M_r. As t increases, this molecule moves in accordance with the differential equations and will sweep out a tube in M which must ultimately overlap itself because of the property of regional recurrence. Let R denote the tubular region so described together with its limit points. As t increases, the end of the tube R moves into the tube, and the region R is carried into all or part of itself. But, because of the property of regional recurrence, this region cannot move into part of itself, and so is carried precisely into itself, as t increases, or decreases. Thus the complete tube formed from the molecule by allowing t to vary in either direction yields the same region R, made up of complete motions.

Now two possibilities arise: either for every point P of M, and any molecule about the point, a region R is obtained which coincides with M, or for some point P and choice of an enclosing molecule, R is only part of M.

In the first case we shall call the connected part of the set of central motions M_r of 'transitive' type, whereas in the second case it is of 'intransitive' type.

For the problem of classical dynamics, transitivity means that any small molecule ultimately sweeps out the entire manifold M of states of motion (except for nowhere dense motions), whereas this is not true in the intransitive case.

A necessary and sufficient condition for the intransitivity of a connected set of central motions is that there exists an invariant closed region in the set, forming only part of it.

The characteristic of the intransitive case of classical dynamics is thus that there exist invariant n-dimensional continua of complete motions filling only part of M.

Evidently the stated condition is necessary since such an invariant region is found in the region R described above. On the other hand, if there is an invariant region R in M_r, a molecule lying within R must always continue to lie in R for increasing or decreasing t, so that the motions are intransitive.

In the intransitive case all motions are special. For an arbitrary motion either lies in such an invariant sub-continuum, or in the complementary set of invariant sub-continua, or on the boundary of the given invariant sub-continuum.

In case a connected part of the set of central motions M_r is transitive, there will exist motions which, as t either increases or decreases, ultimately pass arbitrarily near all points of the manifold of states of motion.

For definiteness we shall take $M_r = M$ in our demonstration. Moreover we make the preliminary observation that every molecule must fill M with increasing t; else it would define an invariant partial region R of M of the type excluded in the transitive case. Hence there exist arcs of curves of motion which go with increasing t from the neighborhood of a given point P to that of a second given point Q.

Let us begin by choosing a positive quantity d less than 1, and any numerable set of points P_k, $(k = 1, 2, \cdots)$, which is everywhere dense in M. It is clear that if a second set $\bar{P}_k$, $(k = 1, 2, \cdots)$ is assigned such that P_k is distant at most d^k from $\bar{P}_k$ for $k = 1, 2, \cdots$, then the second set will also be everywhere dense in M.

To obtain a motion which is not special we may proceed in the following manner. Within the d neighborhood of P_1, the point P_1' and an arc $P_1' \, P_2'$ of a curve of motion can be found such that P_2' lies in the d^2 neighborhood of P_2. Mark now about P_1' a smaller neighborhood, lying within the d neighborhood of P_1 and such that if P_1'' varies anywhere within it, the point P_2'' of the curve of motion through P_1''

still may be taken to vary continuously within the d^2 neighborhood of P_2. This is obviously possible.

Now within this smaller neighborhood of P_1' a point P_1'' may be selected so that the point Q_2'' of the arc of a curve of motion $Q_2'' P_1''$ lies in the d^2 neighborhood of P_2. Thus an arc of a curve of motion $Q_2'' P_1'' P_2''$ is obtained, such that Q_2'' is within the d^2 neighborhood of P_2 while P_1'', P_2'' are within the d and d^2 neighborhoods of P_1 and P_2 respectively. Furthermore, we can take a neighborhood of P_1 still in the d neighborhood of P_1, so small that as P_1''' varies continuously within this neighborhood, both P_2''' and Q_2''' of an arc $Q_2''' P_1''' P_2'''$ vary continuously within the d^2 neighborhood of P_2.

By another similar step we may fix upon a P_3''' of an arc $Q_2''' P_1''' P_2''' P_3'''$ so that P_3''' lies within the d^3 neighborhood of P_3. By still another step we obtain an arc $Q_3^{(4)} \cdots P_3^{(4)}$, and so we may proceed indefinitely. In this way we construct at the k-th stage an arc of a curve of motion

$$Q_{k-1}^{(k)} \cdots P_1^{(k)} P_2^{(k)} \cdots P_{k-1}^{(k)}$$

such that $P_k^{(j)}$ and $Q_k^{(j)}$ lie in the d^k neighborhood of P_k.

It is clear that by a limiting process we arrive at a curve of motion

$$\cdots Q_2^* \, Q_1^* \, P_1^* \, P_2^* \, P_3^* \cdots$$

in which P_k^* and Q_k^* lie within the d^k neighborhood of P_k. Consequently the sets $P_1^*, P_2^*, \cdots$ and $Q_1^*, Q_2^*, \cdots$ are everywhere dense in M. Hence the α limit and the ω limit motions make up all of M, and the motion itself is not special.

In the following chapter (section 11) an example of a nonsingular geodesic problem of transitive type is given. It seems probable that, in general, after the obvious reductions by means of known integrals are effected, the problems of classical dynamics are of transitive type.

Between the general transitive case and the highly specialized cases of completely integrable type, there is a prodigious variety of intermediate possibilities, dependent on the particular properties of the differential equations.

In the following chapter we consider the case of a system with two degrees of freedom. Unfortunately it does not seem to be the fact that the methods there employed admit of simple extension to the case of more degrees of freedom. The problem of three bodies, treated in chapter IX, is extremely instructive as an instance of this more complicated case, although it is of singular type.

CHAPTER VIII

THE CASE OF TWO DEGREES OF FREEDOM

1. Formal classification of periodic motions. In chapter VI we studied dynamical systems of Hamiltonian type ($m = 2$) in a preliminary way, with particular reference to the periodic motions. We propose now not only to obtain a more complete idea as to the existence and distribution of these periodic motions, but also of the various other types of motion.

For systems of this type the manifold of states of motion is four-dimensional at the outset, with coördinates p_1, q_1, p_2, q_2. However, by specification of the constant of energy, $H = h$, a three-dimensional analytic sub-manifold is defined, and it is a particular such manifold M which is the manifold of states of motion under consideration. In other words, by use of the energy integral the system of differential equations is reduced from the fourth to the third order.

In order to limit attention to a definite type of case we assume that M is non-singular, i. e. closed and analytic, and furthermore we exclude the possibility of equilibrium in M, since equilibrium cannot arise for a general value of h.

Consider now a periodic motion, which will be represented by a closed curve in M. Imagine that curve to be cut by an analytic surface S. If a point P of S is followed along its corresponding curve of motion in the sense of increasing time, it will intersect S again at a point P_1. We write $P_1 = T(P)$, thereby defining a one-to-one, analytic transformation of S into itself, at least in the neighborhood of the given periodic motion. The transformation T leaves invariant the point corresponding to the periodic motion. For periodic motions near to the given periodic motion, but represented by curves making k circuits before closing, there will be a corresponding set of k points, $P, T(P), \cdots, T^{k-1}(P)$ with $T^k(P) = P$, where the meaning of the notation is manifest.

There is a particular and important case in which we can specify the characteristic properties of the transformation T, on the basis of our earlier work. This is the case in which the Hamiltonian problem is derived from a Lagrangian problem with principal function quadratic in the velocities (chapter VI, sections 1–3).

Let us recall the precise method of selecting the coördinates in this case. In the first place the Lagrangian coördinates q_1, q_2 are so selected that q_1 is an angular coördinate reducing to $2\pi t/\tau$ (τ, the period) along the periodic motion, while q_2 is 0 along it. Then, from the differential equations, $\partial H/\partial p_1$ is not 0 along the periodic motion, and we can solve the equation $H = h$ for p_1 in the form

$$p_1 + K(q_1, p_2, q_2, h) = 0,$$

where K is analytic in its four arguments and is periodic of period 2π in q_1. Hence q_1, p_2, q_2 constitute a suitable set of coördinates for M in the torus-shaped vicinity of the given periodic motion. In these variables the equations of Hamiltonian type

$$\frac{dp_2}{dq_1} = -\frac{\partial K}{\partial q_2}, \quad \frac{dq_2}{dq_1} = \frac{\partial K}{\partial p_2} \tag{1}$$

subsist. The equations (1) enable us to express the coördinates p_2, q_2 of any curve of motion in terms of the angular coördinate q_1; and t may then be found by a simple integration.

A final modification is to use $p_2 - \overset{0}{p_2}$ as coördinate instead of p_2. Here $\overset{0}{p_2}(t)$ is the expression for p_2 along the given curve where t is thought of as replaced by its value $q_1 \tau/2\pi$ along the periodic motion. If at the same time we modify K by adding a term $q_2\, d\overset{0}{p_2}/dq_1$, the form (1) is preserved, and the periodic motion corresponds to $p_2 = q_2 = 0$. Furthermore K remains periodic of period 2π in q_1.

In this way the nature of the coördinates employed in the reduction to generalized equilibrium becomes apparent. These coördinates have the advantage that when the 'plane' $q_1 = 0$ is taken as the surface of section S, the transformation T becomes area-preserving. Furthermore, if the periodic motion is of general stable type, so that the multipliers $\pm\lambda$ in (1) are pure imaginary and incommensurable with $\sqrt{-1}$, it was seen (loc. cit.) that T may be given the normal form

$$(2)\quad \begin{aligned} u_1 &= u_0 \cos(\sigma + s r_0^2) - v_0 \sin(\sigma + s r_0^2) + \Phi \\ v_1 &= u_0 \sin(\sigma + s r_0^2) + v_0 \cos(\sigma + s r_0^2) + \Psi \end{aligned} \qquad (r_0^2 = u_0^2 + v_0^2),$$

in suitably chosen variables u, v, where Φ, Ψ are given by convergent power series in u_0, v_0 starting off with terms of arbitrarily high degree, at least if a certain quantity $l = \sqrt{-1}\, s/2\pi$ does not vanish.

It should be remarked that the choice of the surface S does not affect the transformation T obtained except by a change of variables.

The detailed study of the transformation T on the basis of the area-preserving property and the normal form (2) enabled us to infer that infinitely many periodic motions exist in the immediate vicinity of the given periodic motion of stable type.

In the case when the periodic motion is of general unstable type the multiplier λ is either real, or $2\lambda - \sqrt{-1}$ may be taken real. The same method as was employed in the stable case (chapter III, sections 6—9) leads to an analogous formal solution, and to a real normal form for T

$$(3)\quad u_1 = \mu u e^{l r_0^2} + \Phi, \qquad v_1 = \frac{1}{\mu} v e^{-l r_0^2} + \Psi \qquad (\mu \neq \pm 1),$$

in the case $l \neq 0$, where Φ, Ψ are of the same type as in (2).

This general unstable case is very simple to treat analytically.* There will be two invariant analytic curves through

* Cf. my paper, *Surface Transformations and Their Dynamical Applications,* Acta Mathematica, vol. 43 (1912), section 27, or the paper by Hadamard, *Sur l'itération et les solutions asymptotiques des équations différentielles,* Bull. Soc. Math. France, vol. 29 (1901).

the origin, which may be taken to be the u and v axes. Points along one of these invariant curves approach the origin upon successive iteration of T; points on the other invariant curve leave the vicinity of the origin; while points not on these curves first approach and then recede from the origin.

If then we interpret this situation in the manifold M near a periodic motion of unstable type, we find that there are two invariant analytic surfaces through the curve of periodic motion, one of these corresponding to an analytic family of positively asymptotic motions, the other to an analytic family of negatively asymptotic motions. Other nearby motions first approach and then recede from the given periodic motion of unstable type.

It is clear that there can be no periodic motions whatsoever lying wholly near to the periodic motion of unstable type, in contrast with the fact that there must be periodic motions near to any periodic motion of general stable type ($l \neq 0$).

In this way it is seen how fundamentally the classes of periodic motions of stable and unstable type differ from one another.

We are now prepared to state to what extent the apparent limitations introduced are necessary.

In the first place not all the first partial derivatives of H can vanish at any point of the periodic motion, so that the manifold $H = h$ is a regular analytic three-dimensional manifold M along the periodic motion. If variables p, q, r, h are selected as coördinates instead of p_1, q_1, p_2, q_2, the invariant integral of ordinary four-dimensional volume takes the form

$$\int\int\int\int \varphi \, dp \, dq \, dr \, dh$$

where $\varphi > 0$ is analytic in p, q, r, h. Hence $\int\int\int \varphi \, dp \, dq \, dr$ is invariant in M.

It follows further that T will leave a double integral $\int\int \psi \, du \, dv$ invariant, where u, v are coördinates in S, and

$\psi > 0$ is analytic in u and v. The argument is essentially that of chapter VI, section 1. This fact alone suffices to lead to the normal forms (2), (3), and to the conclusions cited above.*

Hence the Hamiltonian problem need not be restricted in this manner.

In fact it is found that for the most general transformation T with such an invariant double integral $\iint \psi \, du \, dv$, there is always a formally invariant function $\Omega(u, v)$ given by a formal power series in u, v. We may define the case in which the equation $\Omega = 0$ yields real formal invariant curves as of unstable type. In this case there are always asymptotic invariant analytic families of motions (or analytic families of periodic motions containing the given periodic motion). All other nearby motions approach and then recede from the given motion. Thus there are no nearby periodic motions, except those that belong to the same analytic family as the given periodic motion, if there are such.

If $\Omega = 0$ yields no real formal invariant curve of this kind, the periodic motion may be called of stable type. In the general stable case treated above, Ω is r^2, to terms of higher order. When σ is incommensurable with 2π, while s, together with some but not all of the set of analogous constants perhaps, vanishes, no essential modification is required except that the term sr_0^2 in (2) is replaced by a term $s^{(k)} r_0^{2k}$. If, however, all of these constants vanish, the normal form (2) holds with $s = 0$, and for these irregular periodic motions it is no longer possible to apply the reasoning by which the existence of infinitely many nearby periodic motions was established.

On the other hand, no essential difficulty arises in the case of stable type when σ is 0 or $\pm \pi$ or, more generally, is commensurable with 2π; this is the case when the given periodic motion is multiple, at least when taken as

* See my paper (loc. cit.) for justification of the fact stated as well as of what follows.

described a certain number, k, of times. It is only necessary to consider T^k in place of T, for which the number σ is also 0. Here the invariant function Ω starts off with higher degree terms than the second, and casual inspection indicates that T is analogous to a rotation through an angle which vanishes at the origin but increases (or decreases) with distance from the origin. It would therefore seem highly probable that in this case too there must be infinitely many neighboring periodic motions, although the analytic details need to be carried through.

Consequently it appears that in very general stable cases, and probably in all cases except the highly exceptional case when T is equivalent formally to a pure rotation through an angle incommensurable with 2π, this property will continue to hold. This exceptional case is that in which the function M in the formal solution reduces to its first term λ.

Hence in the most general case of unstable type ($m = 2$) *the phenomenon of asymptotic analytic families of motions* (*or at least of analytic families of periodic motions containing the given motion*) *is characteristic. Other nearby motions approach and then recede from the given periodic motion.*

In the most general stable case, except the highly degenerate case where σ is incommensurable with 2π and the formal series involve no variable periods, there will be neighboring periodic motions.

It is to be emphasized that the second of these conclusions has been formulated without completion of a detailed proof, such as I have not yet had the opportunity to effect.

The degenerate case of stable type includes a real exception, as the example of chapter VI, section 4, shows, and should be further studied. Moreover the formal series break down in the stable case when λ is commensurable with $\sqrt{-1}$. These must be replaced by much more complicated types of series, a suggestion for the structure of which may perhaps be found in my paper referred to above; the earlier definition of complete formal stability will need to be extended so as to permit of indefinitely large periods.

Between the non-specialized dynamical problem and the highly exceptional integrable case, there exists an enormous variety of intermediate cases. In order to possess the analytical weapons with which to treat all cases whatsoever, it will undoubtedly be necessary to treat the question of the stability and instability of analytic families of periodic motions in much the same way as that which is outlined for the periodic motions above. While the individual periodic motions in such a family are to be regarded as unstable, this fact yields no information as to how nearby motions behave with respect to the family of motions as a whole.

In order to avoid complication then, rather than because of any essential mathematical difficulty, we propose to deal mainly with the class of dynamical problems for which every periodic motion and its multiples are simple with $l \neq 0$. Such systems will be termed 'non-integrable systems of general type'. The integrable case will be treated separately (section 13), while indications as to the nature of the result in the intermediate cases will be given.

2. Distribution of periodic motions of stable type. Our first aim will be to establish the following result:

For non-integrable Hamiltonian systems of general type ($m = 2$), *the set of periodic motions of general stable type is dense on itself in* M.

It will be observed that this result constitutes a slight improvement over the result of chapter VI, sections 1–3, according to which other periodic motions, stable or unstable, lie near such a periodic motion of stable type.

To begin with, we recall the facts developed in the lemma of chapter VI, section 1. It was found there that an arbitrarily small vicinity of the origin, $r \leqq \varrho$ (r, θ, polar coördinates) can be selected at pleasure, and then an integer n such that (1) all the points of $r \leqq \varrho$ remain in the region $r \leqq 2\varrho$ under $T, T^2, \cdots, T^n$ and (2) $\partial \theta_n / \partial r_0$ is positive for $r \leqq \varrho$, θ_n being at least 2π greater for $r = \varrho$ than for $r = 0$. It is easy to extend the argument to show that $\partial r_n / \partial r_0$, $\partial \theta_n / \partial \theta_0$ are positive under the same circumstances.

In this way the curve

$$\theta_n - \theta_0 - 2k\pi = 0,$$

where k is so chosen that the left-hand member is negative but not less than -2π for $r = 0$, will have one and only one point (r, θ) on each radius vector, with $r \leqq \varrho$. Thus the equation written defines an analytic curve C encircling the origin and meeting each radius vector only once. But, by the defining property of C, each point P of C goes into a point P_n on the same radius vector; thus the curve C_n is also met only once by any radius vector. Also, because of the area-preserving property of T^n, C_n and C will intersect in at least two points, and these are obviously invariant points of T^n. In the case under consideration C_n and C cannot coincide, for C would then correspond to an analytic family of multiple periodic motions.

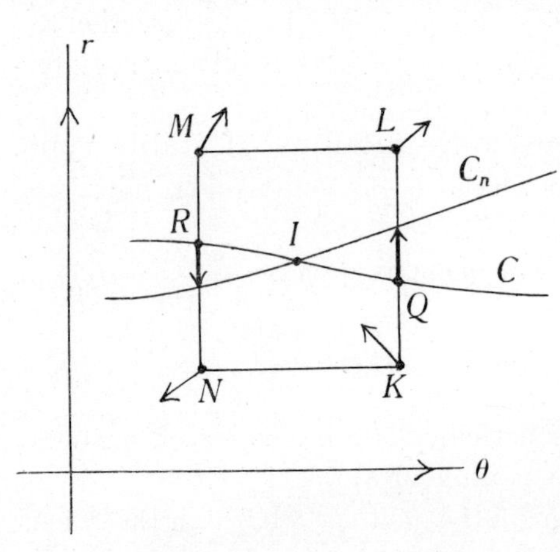

We propose to consider more closely the indices of these invariant points. Let us regard r, θ as rectangular coördinates and consider the adjoining figure in which I is an invariant point at which the curve C_n passes from within C to outside of C, as a moving point describes C in the sense of increasing θ.

If a point P makes a positive circuit of I, for instance around a rectangle $KLMN$, the vector PP_n will have a component to the right above C, a component to the left below C, as follows from the facts noted above. At the points Q and R the vector PP_n is directed upwards and downwards respectively. It is therefore apparent that during the circuit, the vector PP_n rotates through an angle $+2\pi$, so that the index of I is $+1$; while the invariant point J, at which C_n crosses C in the opposite sense has an index -1.

Now by hypothesis the periodic motions corresponding to I and J are not multiple. If of stable type, the number σ is not commensurable with 2π. The corresponding normal forms are either of the type (3), in which μ is positive or negative but not ± 1, or of the type (2).

By the aid of these forms it is easy to determine the respective indices. In the first and second cases, the slope of the vector PP_n is

$$\frac{v_n - v}{u_n - u} = -\frac{1}{\mu}\frac{v + \cdots}{u + \cdots},$$

where only the first order terms are indicated explicitly in the numerator and denominator. It thus appears that if μ is negative the rotation is the same as that of the vector drawn from the invariant point I at the origin to the point (u, v), i. e., 2π. Hence the index is $+1$ if μ is negative; likewise the index is obviously -1 if μ is positive. Moreover, in the third case of stable type when the number σ is by hypothesis incommensurable with 2π in (2), T^n is approximately a rotation through an angle incommensurable with 2π near the origin, so that the vector PP_n rotates through 2π during such a circuit. Hence the index is $+1$ if the motion is of stable type.

We infer then that J corresponds to a periodic motion of unstable type, while it is not clear as yet whether I is of stable or unstable type.

As a matter of fact, however, under the conditions stated I must be stable. The numbers μ are the roots of the characteristic equation, which takes the form

$$\begin{vmatrix} \dfrac{\partial r_n}{\partial r_0} - \mu, & \dfrac{\partial r_n}{\partial \theta_0} \\ \dfrac{\partial \theta_n}{\partial r_0}, & \dfrac{\partial \theta_n}{\partial \theta_0} - \mu \end{vmatrix} = 0$$

when the variables r, θ are used. Since the roots are the reciprocals of one another, this equation reduces to

$$\mu^2 - \left(\frac{\partial r_n}{\partial r_0} + \frac{\partial \theta_n}{\partial \theta_0}\right)\mu + 1 = 0,$$

in which we are assured that the coefficient of μ is negative. Consequently μ is positive and I corresponds to a periodic motion of stable type.

This disposes of the general case when there are no multiple periodic motions and $l \neq 0$.

If the original motion is of stable type but not of that highly exceptional type when there are no variable periods in the formal series, it seems to me that analogous results are to be expected, i. e. that there will exist nearby periodic motions of stable type.

This exceptional case merits particular attention; it is conceivable that it can only arise for integrable dynamical problems.

3. Distribution of quasi-periodic motions. Let us suppose that there exists at least one periodic motion of stable type for the Hamiltonian system under consideration, taken of non-integrable general type. This motion is represented by a closed curve C in the manifold M of states of motion.

Now select any such closed curve C_1 of motion of stable type. Very near to it can be found a closed curve C_2 of motion of stable type which makes k_1 circuits of C before closing. Next choose a closed curve of motion C_3 of stable type very near to C_2 and making k_2 circuits of C_2, and so $k_1 k_2$ circuits of C_1, before closing. Thus we obtain a sequence of closed curves C_n, $(n = 1, 2, \cdots)$, which can evidently be chosen so as to tend toward a definite geometric limiting set C as n becomes infinite, merely by restricting sufficiently the successive neighborhoods of C_1, C_2, $\cdots$. Furthermore, we can prevent C from being itself one of the numerable set of closed curves of motion by the same process. For instance at the nth stage we might confine attention to a neighborhood of C_n which is so small as to contain no closed curve of motion (other than C_n) of length less than n; there are only a finite number of such motions of course.

It is interesting to inquire into the analytic form of the set C. Let q_1 be the angular coördinate in M which increases by 2π when a single circuit of C is made. Then p_1, p_2, q_2 may be thought of as appropriate coördinates of this motion (section 1), and we may write

$$p_1 = f_1(q_1), \quad p_2 = g_1(q_1), \quad q_2 = h_1(q_1), \quad t = \int k_1(q_1)\,dq_1$$

as the equations of the periodic motion C_1, in which $f_1, g_1, h_1, k_1 > 0$ are analytic periodic functions of q_1 of periodic 2π. For C_2 we have likewise

$$p_2 = f_2(q_1), \quad p_2 = g_2(q_1), \quad q_2 = h_2(q_1), \quad t = \int k_2(q_1)\,dq_1$$

where f_2, g_2, h_2, k_2, are analytic periodic functions of q_1 of period $2k_1\pi$. Thus we form in succession a sequence of functions f_n, g_n, h_n, k_n, periodic of period $2k_1 \cdots k_{n-1}\pi$ in q_1, corresponding to the periodic motions C_n, $(n = 1, 2, \cdots)$. If we take points $q_1 = 0$ so as to approach a limit, it is obvious that f_n, g_n, h_n, θ_n approach limits f, g, h, θ, where the limits are approached uniformly for all values of q_1.

If there exists a single periodic motion of stable type for a non-integrable Hamiltonian problem of general type, there will exist infinitely many nearby motions, quasi-periodic but not periodic, with coördinates of the form

$$p_1 = \lim_{n \doteq \infty} f_n(q_1), \quad p_2 = \lim_{n \doteq \infty} g_n(q_1),$$
$$q_2 = \lim_{n \doteq \infty} h_n(q_1), \quad t = \int \lim_{n \doteq \infty} k_n(q_1)\,dq_1,$$

where f_n, g_n, h_n, k_n are analytic periodic functions of q_1 with periods $2\pi k_1 \cdots k_{n-1}$, $k_1, \cdots, k_n$ being positive integers which may be taken greater than 1. *The convergence is uniform for all values of q_1.*

Evidently there is a non-denumerable number of such quasi-periodic motions, and the coördinates are functions of the type treated by Bohr. It is clear that they constitute a class of recurrent motions of a new type.

4. Stability and instability. For the consideration of the periodic motions of stable type in dynamical problems, a fundamental division of cases must be made. It may happen that all motions sufficiently near the given periodic motion remain in a small neighborhood for all time. This is the simpler of the two cases, in which case the periodic motion in question may be termed 'stable'. The other possibility is that for some small fixed neighborhood of the given periodic motion, there may be found motions which are arbitrarily near the given periodic motion at the outset but ultimately pass out of the fixed neighborhood. In this case the periodic motion in question may be termed 'unstable'.

Evidently the classification here effected may be made not only for periodic motions but also for recurrent motions of any type. Stability in this fundamental qualitative sense is not to be confused with the 'complete formal stability' introduced earlier, and a periodic motion 'of stable type' may or may not be stable.

The transformation T of the surface S yields an immediate simple condition for stability.

Consider a small region s of S about the invariant point, and its images $s_1, s_2, \cdots$ under successive applications of the transformation T. All of these contain the invariant point as interior point. The infinite set of regions $s, s_1, \cdots$ will lie in the vicinity of the invariant point, according to the hypothesis of stability. These regions taken together occlude a certain neighborhood $\bar{s}$ of the origin, which is taken into all or part of itself by the transformation T since the set $s, s_1, s_2, \cdots$ is taken into $s_1, s_2, \cdots$. But $\bar{s}$ cannot be taken into a part of itself because of the existence of an invariant area integral. Hence $\bar{s}$ yields an invariant area in S, corresponding to which there is an invariant torus-shaped region of M enclosing the curve of the given periodic motion in its interior.

A necessary and sufficient condition for stability is the existence of infinitely many invariant torus-shaped regions closing down upon the curve of the given periodic motion in the manifold M of states of motion.

5. The stable case. *Zones of instability.* What then is the nature of the boundary of such an invariant torus-shaped region in M surrounding the given closed curve of stable periodic motion? In answering this question we naturally turn to consider the nature of the corresponding invariant closed curve in S forming the outer boundary of invariant region $\bar{s}$.

Let us assume that the periodic motion, although of general stable type is not such that the formal series involve no variable periods (section 1). In this case a normal form (2) with $s \neq 0$ (or a similar form) can be used. We may assume that s is positive, for if s is negative for T the corresponding quantity, $-s$, is positive for T^{-1}. This normal form (2) shows that the counter-clockwise rotation about the invariant point increases with radial distance if r is sufficiently small.

No region $\bar{s}$ lying very near to the invariant point can be met more than once by some radial line. In fact let s^* denote the part of the plane formed by the radial lines extended to the most distant points on the boundary of $\bar{s}$. The regions of s^* not forming part of $\bar{s}$ are of one of two possible types: either they are bounded by the boundary of $\bar{s}$ and a piece of a radial line on the left, or by the boundary of $\bar{s}$ and a piece of a radial line on the right. But the transformation T evidently takes a region bounded on the left by such a radial line l into a region bounded by $\bar{s}$ and the image l_1 of this radial line l. Since the angular coördinate increases with the radius, it is geometrically evident (see figure) that

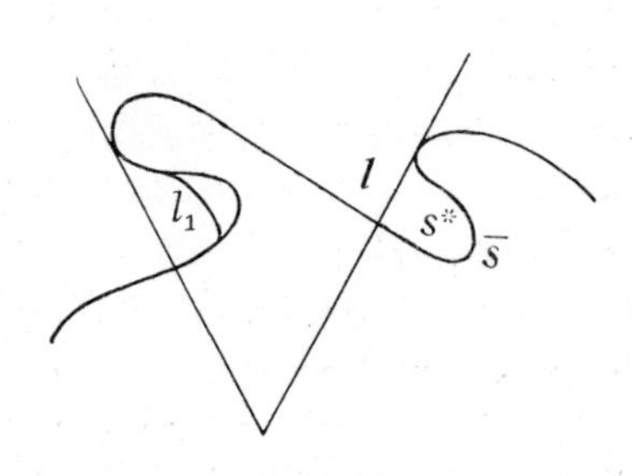

the image of such a region will necessarily fall within a region of the same type.

But this situation is not possible since the transformation admits an invariant area integral, and consequently no set of regions can be taken into part of itself by T. Thus there exist no regions bounded on the left by such radial lines.

Similarly the use of the inverse transformation T^{-1} shows that there are no such regions bounded on the right by such radial lines.

Hence the boundary of the invariant region $\bar{s}$ is met only once by any radius sector.

The impossibility of a radial segment forming part of the boundary of $\bar{s}$ is obvious, so that there is actually only one point of intersection with each radial line.

A more elaborate consideration based on the normal form shows that the boundary curve $r = f(\theta)$ is one for which the difference quotient

$$[f(\theta_1) - f(\theta_2)]/(\theta_1 - \theta_2)$$

is bounded and indeed small for invariant regions near enough to the invariant point.* The truth of the fact seems almost obvious if one observes that T rotates positively directions differing by any considerable amount from the directions perpendicular to the radial direction into other directions differing still more from that direction.

Our conclusion may thus be summarized in the form:

For a stable periodic motion of general stable type and with variable periods in the formal series, the invariant torus-shaped regions are such that their intersections with the analytic surface of section S may be represented in the form $r = f(\theta)$, where r, θ are polar coördinates with the invariant point at the origin, and where f is a continuous periodic function of θ of period 2π for which the difference quotient is bounded.

The curves of motion on the boundary of such a torus-shaped region form a closed invariant family. In any such

* For details see my paper (loc. cit.), sections 42–48.

closed invariant family of motions near the given stable periodic motion, there is of course at least one recurrent motion.

If the rotation number $\varkappa$ along the corresponding invariant curve on S is incommensurable with 2π the surface of the torus may represent a single minimal set of recurrent motions. In this case the coördinates and the time can be expressed in terms of continuous doubly periodic functions. In order to make this clear, let us first select angular coördinates θ, φ on the torus as follows. The coördinate φ will be taken to vanish in S, and to increase proportionately with the time along each curve of motion, the factor of proportionality being so taken as to increase φ by 2π between successive intersections with S. The coördinate θ will be defined along the invariant curve on S so that the transformation T takes the form $\theta_1 = \theta + \varkappa$ where $\varkappa$ is the rotation number specified. Elsewhere on the torus the variable θ may be defined as the θ of the corresponding point on S diminished by $\varkappa\varphi/2\pi$ in order to make θ single-valued on the torus. In these coördinates the equation of a curve of motion is $\theta - \theta_0 = \varkappa\varphi/2\pi$. Moreover the coördinates p_1, q_1, p_2, q_2 are doubly-periodic continuous functions of θ, φ, and $dt/d\varphi$ is also. Hence we may write

$$p_1 = f(\varkappa\varphi/2\pi, \varphi), \quad q_1 = g(\varkappa\varphi/2\pi, \varphi), \quad p_2 = h(\varkappa\varphi/2\pi, \varphi),$$
$$q_2 = k(\varkappa\varphi/2\pi, \varphi), \quad t = \int l(\varkappa\varphi/2\pi, \varphi)\, d\varphi,$$

where f, g, h, k, l are continuous doubly periodic functions of period 2π in their two arguments.

There is a second possibility to be considered also. The minimal set of curves of motion may correspond to a perfect nowhere dense set of points on the invariant curve; all other curves of motion on the surface of the torus-shaped region will then approach this minimal set of recurrent motions asymptotically as the time t either increases or decreases.*

* For proof of these facts and reference to the prior work of Poincaré, see my paper *Quelques théorèmes générales sur le mouvement des systèmes dynamiques*, Bull. Soc. Math. France, vol. 40 (1912).

When, however, the rotation number is not incommensurable with 2π, but is $2p\pi/q$ (p, q, relatively prime integers), there will of necessity exist points of the invariant curve which are invariant under T^q. It can be proved that the entire curve is then made up of analytic arcs terminated by points invariant under T^q, while the interior points of such arcs tend asymptotically towards these invariant points upon iteration of T or its inverse.* We are thus led to the following conclusion.

Any such closed invariant family of motions near the given stable periodic motion of general stable type and with variable periods in the formal series is characterized by a rotation number. If this number is incommensurable with 2π, either the family consists of a single minimal set of recurrent motions of continuous type, or it contains a perfect nowhere dense minimal set of recurrent motions of discontiuous type which all other motions of the family approach asymptotically as t increases or decreases. If this number is commensurable with 2π, there exists one or more closed periodic motions in the family, while the other motions form analytic branches asymptotic to these periodic motions.

It may be observed that this is a result concerning invariant sub-manifolds of the manifold M and, in particular, concerning the central motions in this sub-manifold. Incidentally it appears that, although in dynamical systems of classical type, all the motions are central with reference to the whole manifold, the same is not necessarily true of invariant sub-manifolds, so that the concept of central motions continues to play a part even in the problems of classical dynamics.

Any two of these closed families must be entirely distinct from one another, except when both have the same rotation number, commensurable with 2π. Clearly the rotation number, which measures the mean angular rotation, must be the same for two intersecting families. To establish that this number must be commensurable with 2π, we note that since the two families have at least one motion in common although they do not coincide, the two corresponding curves, $r = f_1(\theta)$,

* See my paper in the Acta Mathematica, loc. cit., sections 42–48.

$r = f_2(\theta)$ in S will enclose one or more areas between them, each bounded by a single arc of either curve. Under iteration of T this area must ultimately overlap itself and so coincide with itself, the two arcs going into themselves of course. The two common end points of the two arcs will then be invariant, and hence the rotation number is commensurable with 2π.

Conversely, any two families with distinct rotation numbers must be entirely distinct, the one further away from the periodic motion having the greater rotation number.

By convention let us consider all of the invariant families with the same rotation number commensurable with 2π, as forming a single family. This is natural since any two of the constituent families must then intersect. The outermost boundary of the corresponding network of invariant curves on S, and the innermost boundary curve cannot be wholly distinct of course, since then they would correspond to distinct rotation numbers. This augmented family is clearly composed of a finite number of periodic motions and of certain analytic families of asymptotic motions, according to the statement made above.

Consider now an infinite expanding or contracting sequence of such invariant families. The sequence evidently defines a limiting invariant family, provided it does not close down upon the invariant point, nor expand beyond the neighborhood of a stable periodic motion to which attention is confined.

These invariant families of motions are entirely distinct from one another, with rotation numbers that increase (or decrease) with the distance from the stable periodic motion, and form a closed sequence.

In case the sequence of invariant families of motions contains a pair of successive members, the region of M within the outer of the two corresponding torus-shaped regions and outside the inner one may be called a 'zone of instability'.* On the surface S the zone corresponds to a

* In section 8 the question of the existence of such zones of instability is briefly considered.

ring-shaped region lying between two successive invariant curves. Such regions will certainly exist unless the invariant families fill up the neighborhood of the stable periodic motion completely, aside from the regions occluded by the invariant families with a rotation number commensurable with 2π.

Many of the methods here employed might be used to develop further details concerning the sequence of invariant families, the zones of instability between them and their relation to nearby periodic motions (see sections 8, 9). We shall merely establish the following property:

In any zone of instability about the given stable periodic motion there exist motions passing from an assigned arbitrarily small vicinity of a motion of either of the bounding invariant families into a like arbitrary assigned vicinity of the other bounding invariant family.

In fact consider the inner boundary of the corresponding ring in S, and a small area which abuts on some arbitrary point of that boundary in the ring. Upon indefinite iteration of T an invariant part of the surface S is defined, made up of the part of it formed by the interior of this inner boundary, together with this boundary, the small region and all of its images. The boundary of the invariant part of S so defined must coincide with the outer boundary of the ring, since the inner and outer boundaries are successive invariant curves. But this means that the images of the small area extend arbitrarily near to the outer invariant curve, which is what we wished to prove.

6. A criterion for stability. It is an easy matter to give an analytic criterion for stability.

Let

$$u_1 = f(u, v), \quad v_1 = g(u, v)$$

be the equations defining the transformation T of the surface S in the vicinity of the invariant point and let $r = F(\theta)$ be the equation of one of the invariant curves in polar coördinates. According to the conclusions reached above, F is then a single-valued continuous fuuction of θ_1 of period

2π and with bounded difference quotient. Since this curve is invariant under T it may equally well be written in the form $r_1 = F(\theta_1)$, where the expressions for r_1, θ_1 in terms of r, θ are to be obtained for the coördinate relations above. If then in the modified equations so obtained we replace r by $F(\theta)$, we are lead to an identity and to the following simple criterion:

In order that a periodic motion of general stable type and with variable periods in the formal series be actually stable, it is necessary and sufficient that a certain related functional equation

$$f^2(F\cos\theta, F\sin\theta) + g^2(F\cos\theta, F\sin\theta) = F^2\left(\tan^{-1}\frac{g(F\cos\theta, F\sin\theta)}{f(F\cos\theta, F\sin\theta)}\right)$$

admits of continuous solutions $F(\theta)$, periodic of period 2π with $|F|$ arbitrarily small, but not 0.

7. The problem of stability. An outstanding question in dynamics is whether or not the complete formal stability of a periodic motion of stable type assures stability in the fundamental qualitative sense defined above.

The analytic criteria which distinguish the stable from the unstable case are exceedingly delicate. There are two types of questions which present themselves here. Does formal stability assure such actual stability? If not, does formal stability assure actual stability in important special cases such as the restricted problem of three bodies?

It appears to be certain that in the general case there is instability, although no proof of this conjecture has been obtained. The second of the above questions is much the more difficult one, and is at bottom arithmetic in character; it may be compared to the question of determining whether or not an assigned number is trancendental.

8. The unstable case. Asymptotic families. Let us turn to a like consideration of unstable periodic motions of general stable type and with variable periods in the formal series. In this case there exist no invariant families of curves of the type present in the stable case, at least if attention be restricted to a sufficiently small neighborhood of the given periodic motion. Corresponding to this fact there will be no invariant curves enclosing the invariant point on the surface S.

In such a region of instability about an unstable periodic motion of stable type, there exist two connected families of motions reaching to the boundary of the region, which remain indefinitely within it as t increases and decreases respectively.

To prove this fact we consider as usual the transformation T of the surface S in the corresponding neighborhood of the invariant point. Let σ be a very small region met by each radius vector once and only once. The images $\sigma_n (n = 1, 2, \cdots)$ of σ under T^n must always contain the invariant point within them, and for some value of n must finally extend to the boundary of S; otherwise they would occlude an invariant region $\bar{\sigma}$, whose boundary would be an invariant curve of the excluded type, according to the argument of section 4. The points of σ_n remain in S for n iterations of T^{-1} at least.

Now take the diameter of σ smaller and smaller. The limiting closed set thereby obtained will be connected with the invariant point and the boundary, and must remain in S under all iterations of T^{-1}. If we had started with T^{-1} instead of T in our reasoning, we would have obtained a second similar set remaining within S under all iterations of T. Obviously these two connected sets of points correspond to two connected families of motions possessing the properties specified.

Let us add the hypothesis that the periodic motion is of general stable type with variable periods in the formal series. In that case it has been observed that the transformation T rotates the tangent directions to a curve in a counter clockwise direction, relatively to the radial direction, except for tangent directions nearly perpendicular to the radial direction at the point.

With this property in mind, let us consider the total set Σ_α of points remaining in S under all iterations of T^{-1}, and connected with the invariant point by points of the same kind. According to what has just been established, the set Σ_α extends to the boundary of S.

Imagine any regular curve AB drawn from a point A on the boundary of S, starting in the inward radial direction at A, and never turning to the right of the radial direction. All the points outside of Σ_α are accessible from the boundary of S along such curves AB which have no point in common with Σ_α (see figure). To prove this 'left-handed accessibility' of Σ_α from the boundary of S, we suppose if possible that there are one or more inaccessible regions (see the region σ^* of the figure), which will evidently be partially bounded by radial segments which they lie to the right of in the inward radial direction. The transformation T^{-1} will evidently take these inaccessible regions into parts of themselves, since it rotates radial directions in the clockwise sense relative to the radial direction. But this yields an impossibility, because of the existence of the invariant area integral.

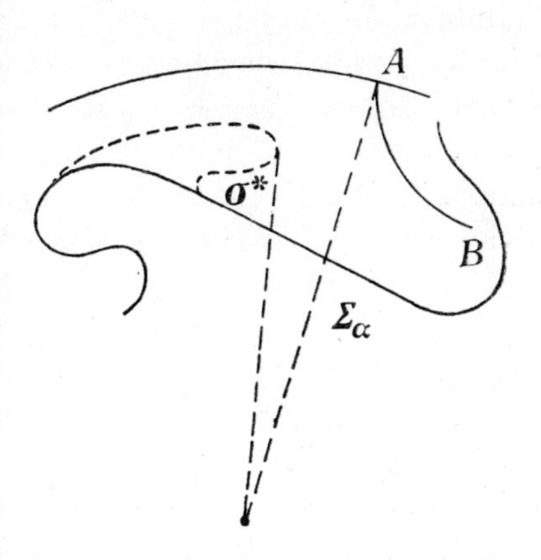

Let the unstable periodic motion be of general stable type with variable periods in the formal series, and suppose for definiteness that the rotation increases away from the motion. Then the closed connected families of motion, Σ_α and Σ_ω, remaining in the region of instability as t decreases or increases respectively and reaching to its boundary, are respectively left-handedly accessible and right-handedly accessible from that boundary.

Let us return to the representation of Σ_α on S. The set Σ_α on S must wind indefinitely often to the right about the invariant point from its intersection with the boundary of S. In order to establish this fact it is convenient to take the polar coördinates θ and r as rectangular coördinates with the θ and r axes directed to the left and upwards respectively. The surface S then appears as an infinite strip, and the region in the strip to the right of and above the connected set Σ_α cannot extend to the right of the point of Σ_α on the boundary of S; otherwise there would clearly be inaccessible regions of the excluded type.

If then Σ_α does not extend indefinitely far to the left, it will be entirely included between two vertical lines in this representation. But the work of section 2 shows that two points, one on $r = 0$ and another nearby, move at sufficiently different rates in the direction of the θ axis so as to separate by an arbitrarily large amount. Hence on sufficient iteration of T^{-1} the curve Σ_α (whose images under T^{-1} all lie in S) must spread over an arbitrarily large strip in the θ direction and so intersect Σ_α. Thus Σ_α and this image will contain an area which must remain in S under all iterations of T^{-1}. But this would lead us to an invariant area $\overline{\sigma}$ in S as before. Hence Σ_α extends indefinitely far in the direction of the negative θ axis. It follows that Σ_α winds indefinitely often to the right about the invariant point, while Σ_ω winds indefinitely often in the opposite sense. Evidently then the two sets Σ_α and Σ_ω intersect infinitely often.

In conclusion we observe that Σ_α must tend uniformly towards the periodic motion under iteration of T^{-1}. Otherwise the limit points of Σ_α under iteration of T^{-1} would yield a closed set Σ'_α, connected with the invariant point and remaining in S under the iteration of T as well as of T^{-1}. The set of such points would then constitute an invariant region $\overline{\sigma}$, whose boundary would be met once and only once by every radius vector according to the argument of section 5. By hypothesis there is no such invariant region.

We may summarize these conclusions in the following way:

The families Σ_α and Σ_ω of motions wind indefinitely often to right or left about the periodic motion, according as they are left-handedly or right-handedly accessible, and so intersect in infinitely many common motions. The motions Σ_α and Σ_ω

are respectively negatively and positively asymptotic to the given periodic motion, while the infinitely many common motions are doubly asymptotic to the given periodic motion.

Entirely analogous considerations to those used above may be applied to any zone of instability. There will exist positively and negatively asymptotic connected sets attached to either boundary and reaching to the vicinity of the opposite boundary. The two sets intersect infinitely often. Furthermore, by the kind of argument employed in the following section, it could be proved that the set positively asymptotic to one boundary intersects the set negatively asymptotic to the other. In consequence there must exist infinitely many motions positively and negatively asymptotic to the two boundaries in any one of the four possible ways.

9. Distribution of motions asymptotic to periodic motions. Hitherto in this chapter we have limited attention to the vicinity of a periodic motion. We turn now to the consideration of the totality of motions in M, which we take to be a closed analytic manifold.

In doing so we shall assume that there exists a surface of section S of genus one, and a corresponding transformation T. The boundaries of S are to correspond to periodic motions of general stable type, and S is cut in the same sense by every curve of motion in M, at least once in any interval of time of sufficient length. Any point P of S if followed along the curve of motion in the sense of increasing time meets it again at P_1, and we write $P_1 = T(P)$, thereby defining a one-to-one analytic transformation of S into itself, which we take to be continuous along the boundaries.

We shall not attempt here to give conditions for the explicit construction of a surface of section S. The details seem to be special to each case (see chapter VI) and not particularly illuminating. Such surfaces of section S and associated transformations T exist in very wide classes of problems.

Moreover we propose to introduce the working hypothesis that the dynamical system is transitive. This hypothesis is certainly satisfied in certain cases as the example given later (section 11) shows, and in all probability is satisfied unless exceptional conditions obtain. However, since the presence of a single stable periodic motion obviously would bring about intransitivity, the conjecture just made can not be established unless the problem of stability is also solved.

If we interpret the hypothesis of transitivity on the surface of section S, it means that, given arbitrary points P_0 and Q_0 of S, then points P and Q arbitrarily near to P_0 and Q_0 respectively and an integer n can be found such that $Q = T^n(P)$.

Suppose now that there exists a single periodic motion of general stable type and with variable periods in the formal series. Since this motion is unstable there exists a network formed by the corresponding connected sets Σ_α and Σ_ω.

Let us consider a boundary of this network which cuts off a small neighborhood of the invariant point. Such a boundary is afforded by the part of S inaccessible from without a given region enclosing the invariant point but without intersection with the corresponding branches Σ_α and Σ_ω. This boundary is not made up wholly of Σ_α or Σ_ω. Otherwise under indefinite iteration of T^{-1} or of T respectively the images of the boundary would continue to lie in that part of S, and an invariant part of S would be defined. Instead it is obvious that under iteration of T^{-1}, for instance, the part Σ_α of the boundary approaches the invariant point, while the part Σ_ω must finally extend into every part of S whatsoever. Otherwise the part of S reached and occluded would also yield an invariant part of S under T, such as is excluded by the hypothesis of transitivity.

Consequently the sets Σ_α and Σ_ω, connected with the invariant point under consideration and asymptotic to it under iteration of T^{-1} and T respectively, are both everywhere dense on the surface of section S.

We have seen (sections 1, 2) that near such a periodic motion of stable type there are infinitely many other periodic motions of stable type which correspond to invariant points of S under some iterate T^n of T. Consider any such periodic motion which is of general stable type with variable periods in the formal series. It has a set Σ'_α and Σ'_ω, both of which must also be dense throughout S.

Now the sets Σ_α and Σ'_α have no points in common, inasmuch as a motion cannot be negatively asymptotic to two different periodic motions. Similarly the sets Σ_ω and Σ'_ω have no points in common.

Hence it is apparent that the sets Σ_α and Σ'_ω have infinitely many points in common, as have the sets Σ_ω and Σ'_α.

In the transitive case of two degrees of freedom when there exists a periodic motion of general stable type having variable periods in the formal series, there will exist infinitely many other periodic motions of stable type. The motions positively or negatively asymptotic to any periodic motion of this infinite set which is of general stable type with variable periods in the formal series, form a set everywhere dense in S. There exist infinitely many motions positively asymptotic to any one of these periodic motions, and at the same time negatively asymptotic to any other periodic motion of the same set, or even to the same periodic motion.

We shall consider next the periodic motions of unstable type and the motions asymptotic to them.

As has been remarked, there exist analytic families of motions positively and negatively asymptotic to such a periodic motion. In the simplest case, to which we may limit attention, there are two corresponding analytic invariant curves through the invariant point, one yielding the positively asymptotic motions and the other the negatively asymptotic motions (section 1). The two arcs of the same invariant curve ending at the invariant point cannot intersect of course, no matter how far extended.

On the contrary, two arcs belonging one to each invariant curve may intersect. In this case the same argument is applicable as was made for the analogous Σ_α, Σ_ω curves above, to show that these two invariant curves are each everywhere dense in S and intersect infinitely often.

Furthermore we can argue as before and obtain the conclusion that infinitely many motions exist, negatively asymptotic to an assigned periodic motion of general stable type with variable periods in the formal series or to an assigned periodic motion of general unstable type with doubly asymptotic motions, and at the same time positively asymptotic to an assigned periodic motion of one of these two types.

It is evident that if two arcs of positively and negatively asymptotic types for such a periodic motion of unstable type intersect the network of one motion of stable type, they will intersect all such networks, and also each other.

If then we can establish that all the four arcs attached to the invariant point intersect these networks it is obvious that we can extend our previous conclusions about motions asymptotic to the periodic motions of stable type to those of unstable type. We shall prove that this is the case provided that (1) the asymptotic analytic arc from one such invariant point of unstable type is not identical to that from a second such invariant point and (2) there is no periodic motion of general stable type with invariable periods in the formal series. The cases when one of these two assumptions fails to hold are to be regarded as highly exceptional. The argument is made only in the case when there are no multiple periodic motions, although the result holds under much wider conditions.

In order to establish this fact, let us suppose that some one of the four arcs belonging to the periodic motion of unstable type does not intersect these networks, and show that a contradiction results.

By extending this arc indefinitely, we obtain a connected limiting set Σ on S. This set Σ is clearly invariant under T. Furthermore, in consequence of this fact and the hypothesis of transitivity Σ cannot enclose an area. Hence, by a well known theorem due to Brouwer, there exists an invariant point of Σ under T. This must correspond to a periodic motion of unstable type, since by hypothesis the extended asymptotic arc cannot approach a periodic motion of stable type with variable periods in the formal series.

Now it is obvious that Σ must contain at least two of the asymptotic arcs of opposite type to this invariant point unless the original extended arc coincides with one of these arcs. But this possibility was also excluded.

It is clear that these new arcs in Σ do not intersect the networks belonging to the periodic motions of stable type, and so we may take any one of them, thus deriving a set Σ_1 within Σ. In this way a process is set up which must finally yield a Σ^* in Σ containing a minimum number of invariant points corresponding to a periodic motion of unstable type, and of asymptotic arcs. Any arc of Σ^* in such a set extended from an invariant point must then have itself as limit point, and there will be at least two such arcs of opposite type to the same invariant point.

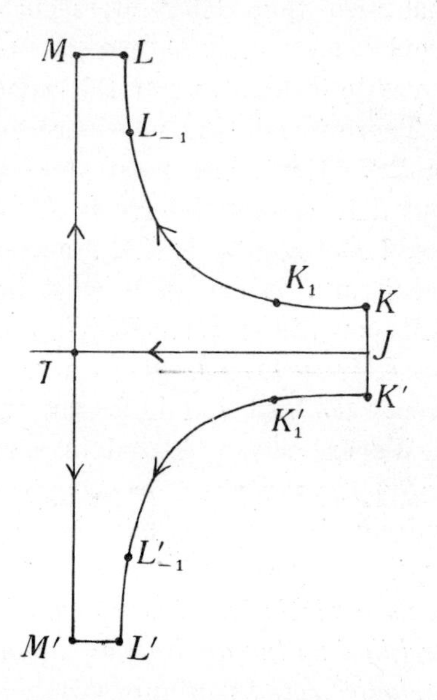

Let us consider separately the cases when there are two, three and four such arcs to the invariant point I (see figure).

To take the first case, let IJ and IM be the positively and negatively asymptotic arcs in Σ^*. Now IM extended must return to the vicinity of I and can only do it along IJ. Let $IJKLM$ be constructed as follows: JK is a short curvilinear arc which meets IJ at J; KL is an arc made up of an arc KK_1 joining K to its image K_1 near to IJ and of its images K_1K_2, K_2K_3, $\cdots$ till the point L is reached near to IM, with LM a short segment. The curvilinear polygon $IJK'L'M'$ is similarly constructed on the opposite side of IJ. The extended arc IM cannot approach IJ along $JK'K_1'$ for then evidently IM' would also lie in Σ^*.

Hence IM produced intersects JKK_1. Likewise IJ produced intersects MLL_{-1}. The *analysis situs* of the figure shows that IM and IJ extended will intersect, contrary to hypothesis.

Next suppose that three arcs, say IJ of one type and IM, IM' of the other, lie in Σ^*. It follows as before that IJ extended meets MLL_{-1} and $M'L'L'_{-1}$. But if there are no intersections of IJ extended with IM or IM', it is apparent from the *analysis situs* of the figure that IM intersects $JK'K'_1$, and also that IM' intersects JKK_1. Thus IM and IM' would necessarily intersect which is impossible since these asymptotic arcs are of the same type.

There remains to consider the case when all four arcs lie in Σ^*. Here IM must intersect $K_1KJK'K'_1$ if extended. But IM cannot intersect $JK'K'_1$, for then IJ if extended could not approach IM', according to the *analysis situs* of the figure. Hence IM must intersect JKK_1, and similarly IM' must intersect $JK'K'_1$. But the same argument applied to IM as to IJ shows that IJ must intersect MLL_{-1}. Hence IJ and IM intersect, which is impossible.

We are now prepared to state our conclusion:

Suppose that there is a surface of section S of genus one for a dynamical problem with two degrees of freedom, of transitive type. Suppose furthermore that all of the periodic motions of general stable type involve variable periods in the formal series, and that no two analytic asymptotic families attached to different periodic motions of unstable type coincide.

Under these circumstance, all of these asymptotic families are dense in M, and there will exist infinitely many motions asymptotic in either sense to any assigned periodic motions, whether of stable or of unstable type.

The special hypothesis made that the periodic motions are not multiple is not essential to the method of attack, but was made in order to restrict attention to the general case, for the sake of simplicity.

The above result makes plain a certain analogy between the motions of stable and unstable type.

It has been proved earlier that in the neighborhood of a periodic motion of general stable type and with variable periods in the formal series, there are nearby periodic motions both of stable and unstable type. Are there also periodic motions which approach arbitrarily near to any periodic motion of unstable type? Of course such a motion cannot remain near throughout a period. The answer is in the affirmative. It may be proved that when the two asymptotic analytic branches of a periodic motion of unstable type intersect, there will be infinitely many periodic motions coming into an arbitrarily small vicinity of the corresponding doubly asymptotic motion and of the given periodic motion of unstable type.*

Thus it appears that in a certain sense the totality of periodic motions, whether of stable or unstable type, will be dense on itself in very general cases. The conjecture of Poincaré that these periodic motions are everywhere dense has been seen not to be always true (chapter VI, section 4), but doubtless holds in very general cases also.

10. On other types of motion. Thus far we have only considered periodic motions, the quasi-periodic motions, and certain other simple types of recurrent motions, among the various types of recurrent motions. Such recurrent motions almost certainly form an endless hierarchy of more and more complicated types, even for non-integrable dynamical systems with two degrees of freedom such as we have been considering. Among motions asymptotic to recurrent motions in either sense we have only considered those asymptotic to periodic motions. We have not considered other special motions, nor non-special motions. The general methods used in this and the preceding chapter yield a variety of results in this connection.†

We shall not attempt to go further in this direction. The

* See my forthcoming paper in the Acta Mathematica, *On the Periodic Motions of Dynamical Systems.*

† See my paper *Surface Transformations and Their Dynamical Applications*, Acta Mathematica, vol. 43 (1922), sections 54–73.

example taken up section 11 will give some idea of the complexity of the situation to be expected.

11. A transitive dynamical problem. The problems of dynamics usually called 'integrable' are those of intransitive type, and the motions are represented by curves which lie upon invariant analytic manifolds in M, of one or two dimensions. For example, in the case of the problem of two bodies, every motion is periodic, and these invariant manifolds in M are closed curves. For the integrable cases (sections 12, 13), the special analytic relations are sufficient to yield complete information about the motions and their interrelations.

Any non-integrable problem of transitive type might, however, be considered to be 'solved', in case a special algorithm of sufficient power could be devised for dealing with it.

I propose here to develop such an algorithm for the transitive geodesic problem on a special analytic surface with negative curvature. The results obtained seem likely to be typical of the general transitive case in many respects, and can be readily extended to any analytic closed surface with negative curvature. It will only be possible to give an intuitive justification for the results. For the technique, I may refer to the notable work of Hadamard* and Morse,† the methods and ideas of which perform the principal role in the case here treated.

It seems improbable that any analogous algorithm exists in the geodesic problem on closed analytic surfaces with positive curvature.

The particular surface which we consider is defined by the equation

$$z^2 = 1 - e^2 \sin^2 \frac{1}{2} x \sin^2 \frac{1}{2} y \qquad (e > 1)$$

* *Les surfaces à courbures opposées et leur lignes géodesiques,* Journ. de Math., ser. 5, vol. 4 (1898).

† *Recurrent Geodesics on a Surface of Negative Curvature,* Trans. Amer. Math. Soc., vol. 22 (1921); *A Fundamental Class of Geodesics on Any Closed Surface of Genus Greater Than One,* Trans. Amer. Math. Soc., vol. 26 (1924).

in which x, y, z are rectangular coördinates, and where we shall make the convention that all points

$$(x \pm 2k\pi,\ y \pm 2l\pi) \qquad (k, l = 0, 1, 2, \cdots)$$

correspond to the same point of the surface. This convention is legitimate inasmuch as the linear group of translations

$$\bar{x} = x \pm 2k\pi, \qquad \bar{y} = y \pm 2l\pi, \qquad \bar{z} = z,$$

takes the surface into itself. A fundamental domain for x, y is then the square given by

$$0 \leqq x < 2\pi, \qquad 0 \leqq y < 2\pi.$$

For any z this equation may be written

$$\sin^2 \frac{1}{2} x \sin^2 \frac{1}{2} y = (1 - z^2)/e^2.$$

Hence for $|z_0| < 1$ the trace of the given surface in the plane $z = z_0$ consists of a convex, symmetrical, analytic oval within the fundamental square, with center at the mid-point of the square. As z_0 increases or decreases from $z_0 = 0$ towards $z_0 = \pm 1$, this oval expands analytically, and becomes the fundamental square for $z_0 = \pm 1$. It is easily verified directly, but is also obvious from the above qualitative considerations, that this surface is everywhere analytic, with curvature which is negative except at the points on the surface corresponding to the corner of the limiting squares, $z = \pm 1$.

The connectivity of this closed surface is readily determined. If the equatorial oval $z = 0$ were capped, the upper half of the surface would be homeomorphic with a square in which pairs of opposite points on parallel sides are identical, i. e. with an anchor ring. When this cap is removed we find that the upper half of the surface is homeomorphic with the surface of an anchor ring having a hole in it. The lower half is of the same type. Thus the total surface is homeomorphic with a double anchor ring of genus 2.

A fundamental property for such a surface of negative curvature is that there is one and one only geodesic arc AB of given type, from the standpoint of *analysis situs,* which joins two assigned points of the surface. For example, take any closed curve circling around one copy of the surface. There is only one closed geodesic of this type, which of course is the oval $z=0$. If we use the complete representation in x, y, z space by means of the surface and the infinitely many congruent copies, the above result means that any continuous line in this surface joining a fixed point A to a fixed point B can be continuously deformed into a unique geodesic arc AB.

Let us seek to introduce a symbolism adequate to express the type of any arc AB. Consider the projection of the given surface upon the equatorial plane. This will cover doubly all of that plane except the part inside any one of the geodesic circles in the equatorial plane. In the adjoining figure these circles are then the images of these geodesic circles, and likewise the horizontal and vertical segments of the square net work having the centers of these circles as vertices, will obviously correspond to closed geodesics.

Suppose that the projection of A lies within a square of the network as well as that of B. As a point P moves from A to B along AB, its projection traces a continuous curve in the x, y plane. Moreover if this projection is given, together with the points on the circles where P moves from a region of the surface $z>0$ to a region $z<0$, then the curve AB is determined.

Let the symbol x denote a crossing of a vertical segment in the positive x direction, and x^{-1} a crossing in the opposite direction; likewise let the symbols y and y^{-1} denote crossings of a horizontal segment in the directions of the positive y axis and the negative y axis respectively. From a point within a square there are four accessible quadrants of circles, in the lower left-hand corner, lower right-hand corner, upper right-hand corner and upper left-hand corner. Denote the corresponding crossings of the geodesic circles toward the positive z axis by w_1, w_2, w_3, w_4, respectively, and also the oppositely directed crossings by $w_1^{-1}, w_2^{-1}, w_3^{-1}, w_4^{-1}$ respectively. It is then geometrically evident that any arc AB corresponds to a symbol formed by a finite set of these 12 symbols written in the same order as the corresponding crossings. Conversely, if such a symbol is given, restricted solely in that after any w_i follows a w_j^{-1}, and vice versa, there will be a unique

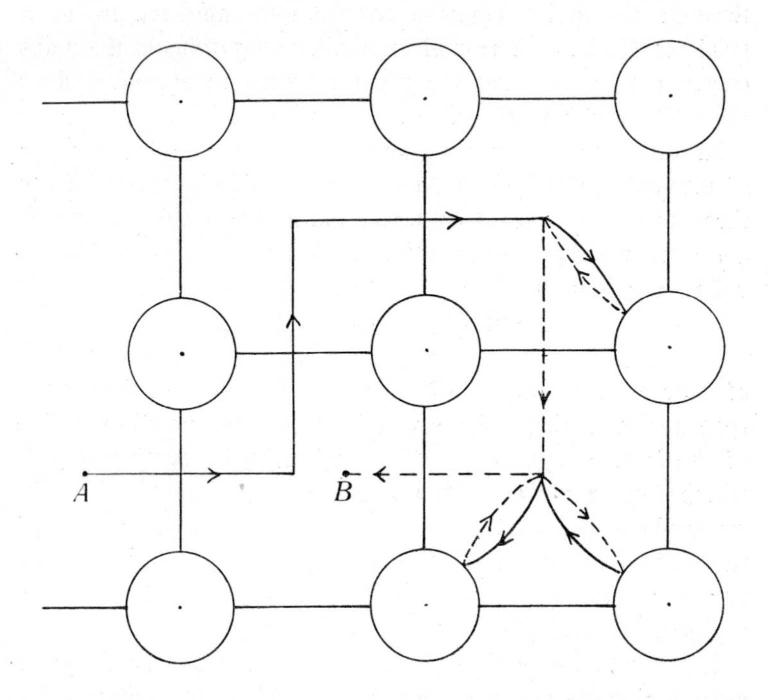

corresponding path. The reason for this restriction is that the point P moves from a region $z<0$ to a region $z>0$, and then from a region $z>0$ to a region $z<0$.

To every allowable deformation of the path AB (A and B being fixed of course) will correspond a modification of the symbol. The class of symbols obtained from one another in this way may be called 'equivalent'. It thus becomes important to determine the legitimate type of modification of a symbol, and a normal form of each class of equivalent symbols.

The legitimate operations are of two types. The first allows us to insert or remove any pair of elements $a\,a^{-1}$ or $a^{-1}\,a$ in the symbol, since this corresponds to a deformation over a boundary. The second operation allows us to replace such a symbol as $w_3\,y$ by $y\,w_2$; or $w_3^{-1}\,y$ by $y\,w_2^{-1}$, or $w_2\,y^{-1}$ by $y^{-1}\,w_3$, or $w_2^{-1}\,y^{-1}$ by $y^{-1}\,w_3^{-1}$; these are the changes possible in the symbol when a point P of AB is deformed through the point common to the quadrants w_2, w_3 of a geodesic circle. There will be similar operations at the point common to w_3, w_4, at the point common to w_4, w_1, and at the point common to w_1, w_2.

In order to obtain a normal form we reduce the number of elements in the symbol as far as possible by the following three processes. First we strike out any pair $a\,a^{-1}$ or $a^{-1}\,a$. Secondly we replace any triple such as $y\,w_2\,y^{-1}$ by its equivalent w_3; here we have

$$y\,w_2\,y^{-1} = w_3\,(y\,y^{-1}) = w_3$$

of course. For each of the sixteen operations of the second type specified above there will be two corresponding triples of the form $p\,w_i\,p^{-1}$ or $p\,w_i^{-1}\,p^{-1}$ where p is x, x^{-1}, y or y^{-1}, which can be replaced by a single letter w_j or w_j^{-1}. Thirdly we replace any triple such as $w_3\,y\,w_2^{-1}$ by y. For each of the sixteen operations of the second type there will also be two corresponding triples of the same sort which can be replaced by a single letter x, x^{-1}, y or y^{-1}.

Lastly whereever possible we invert all ordered pairs such as $y^{-1}\,w_2$ made up of an element x, x^{-1}, y, y^{-1} followed by element w_i or w_i^{-1}, so that an element of the second type appears first. Thus $y^{-1}\,w_2$ is replaced by $w_3\,y^{-1}$.

The normal form will be defined as that which results when all of these processes have been carried out. We propose to show that the normal form is unique. For this purpose we divide the symbol into components made up of the elements x, x^{-1}, y, y^{-1}, and into those made up of the letters w_i, w_i^{-1}. For the arc AB of the figure the symbol is evidently

$$x\,y\,x\,w_2\,y^{-1}\,w_2^{-1}\,w_1\,x^{-1}.$$

It is in normal form and breaks up into the four components

$$x\,y\,x,\ w_2,\ y^{-1},\ w_2^{-1}\,w_1,\ x_3^{-1}.$$

We will argue that this particular normal form for the symbol of AB is unique, but the method used is obviously general. The first component evidently gives the least number of squares passed by the projection of a point P moving from A to B along a path of this type, before the point P crosses the equatorial plane, and it identifies them uniquely in proper order. Hence any other normal symbol for AB must have the same first component xyx as this one. Similarly the second component describes uniquely the greatest number of passages which P can make across $z = 0$ in the same copy of the surface always cutting distinct geodesic circles in succession. In this case there is only one element w_2 in the second component. Thus the second component is w_2 in any other normal symbol.

In general, only the inevitable crossings are made of the sides of the squares and of the geodesic circles, while the latter are made as soon as possible. This is the geometric interpretation of the operations reducing to normal form, and may be made the basis of the proof of uniqueness.

The condition that a symbol is in this normal form merely requires that certain sequences of elements are not to occur. The forbidden sequences are obviously

$$\begin{array}{llll} x\,x^{-1},\ x^{-1}\,x, & y\,y^{-1},\ y^{-1}\,y, & w_i\,w_i^{-1},\ w_i^{-1}\,w_i & (i = 1, 2, 3, 4), \\ x\,w_1,\ x\,w_4, & x\,w_1^{-1},\ x\,w_4^{-1}, & x^{-1}\,w_2,\ x^{-1}\,w_3, & x^{-1}\,w_2^{-1},\ x^{-1}\,w_3^{-1}, \\ y\,w_1,\ y\,w_2, & y\,w_1^{-1},\ y\,w_2^{-1}, & y^{-1}\,w_3,\ y^{-1}\,w_4, & y^{-1}\,w_3^{-1},\ y^{-1}\,w_4^{-1}. \end{array}$$

The infinite symbol corresponding to a complete geodesic will obviously possess the same minimum property as the normal symbol, since the geodesic arc will cross the auxiliary geodesics a minimum number of times. We will call this symbol the 'reduced symbol'. The letters x, x^{-1}, y, y^{-1} follow each other in the reduced symbol ordered precisely as in the normal symbol. But the exact position of the elements w_i can not be determined without a more explicit knowledge of the given surface, and would be different for other surfaces of the same general type.

For the reduced symbol the points of crossing may be associated with the corresponding elements of the symbol, and intermediate points may be indicated by placing an ordinary real number u, $(0 \leqq u \leqq 1)$, between two successive elements α, β, thereby indicating that the point P_u lies the fractional part u of the way from the crossing α to the crossing β along the geodesic arc.

In this way we obtain a symbol for a 'state of motion' in the problem, by adjoining the number u to the complete reduced symbol. Continuous variation of the state of motion will vary u continuously if u be regarded as periodic of period 1. The corresponding variation of the symbol itself is 'continuous' in that a slight variation of the state of motion can produce only changes in a distant element of the reduced symbol, or else will introduce allowable changes of order in successive elements. The corresponding variation in the normal symbol is in general continuous also, but here a component of the form $w_i\, c\, c\, c \cdots$ where c is a given set of elements is to be regarded as the same as $c\, c\, c \cdots$. The two ends evidently vary continuously, but no other changes of order may occur anywhere in the symbol.

We are now prepared to take up the question of the types of motion and their interrelation.

In the first place the periodic motions, corresponding to the closed geodesics, are evidently in one-to-one correspondence with the normal types of finite symbols, in which cyclic order is not considered. The two periodic motions corresponding to the senses in which one and the same closed geodesic may be traversed will correspond to such a symbol and the same symbol taken in the reversed order. The normal symbol for the complete geodesic is evidently the partial symbol taken an infinite number of times. Since the partial symbol may be arbitrarily chosen, the corresponding periodic motion can be made to approach an arbitrary geodesic motion.

The periodic motions are everywhere dense in the totality of motions.

Next let us consider the motions which are positively asymptotic to a given periodic motion. For this to happen, the normal symbol of the motion under consideration must of course be the symbol of the periodic motion from a certain point on, and this condition is sufficient. In order to be negatively asymptotic to the periodic motion generated by the finite symbol p, and positively asymptotic to that generated by q, the symbol must repeat the partial symbol p sufficiently far to the left and q sufficiently far to the right. The intermediate part of the symbol can be taken arbitrarily. If the symbols p and q are the same, a motion doubly asymptotic to one and the same periodic motion may be defined. Since the intermediate part of the symbol is arbitrary the motions of either type are everywhere dense, but must be numerable.

This reasoning shows the existence of a continuous family of positively or negatively asymptotic motions to a given periodic motion. There exist also infinitely many periodic motions positively asymptotic to one assigned periodic motion and negatively asymptotic to a second assigned periodic motion, and these motions are everywhere dense although numerable.

More generally, if a motion is merely to be asymptotic in one sense to any given motion, only one end of the symbol is assigned.

A similar conclusion holds then concerning the motions asymptotic to any two given motions in assigned senses.

There exist also motions semi-asymptotic to the given motions, so that while not actually asymptotic to them the deviations become increasingly infrequent as the time increases or decreases indefinitely.

To make up the corresponding symbol we need merely write down a normal symbol such that in one direction it is in increasing proportion made up of increasing components of the symbol of one of the given motions taken further and further out in its symbol, while in the opposite direction it is similarly made up principally of components of the symbol of the other given motion.

Let us next pass to the general recurrent motion not of periodic type. Evidently the corresponding normal symbol is characterized by the fact that given an arbitrary positive integer n, there exists a second integer N so great that every sequence of n figures in the symbol can be found at least once in any N successive figures.

Morse (loc. cit.) has given a specific method of construction of such a symbol.

There is probably a hierarchy of such recurrent motions dependent in degree of complication upon the way in which N varies with n. Here I wish merely to indicate a possible method which will lead to the discovery of recurrent motions not of periodic type for the system under consideration. Let $f(x_1, \cdots, x_p)$ be any function which is analytic and periodic of period 1 in $x_1, \cdots, x_p$, $(p > 1)$. If $c_1, \cdots, c_p$ are p quantities without linear relation of commensurability between them, then $f(c_1\lambda, \cdots, c_p\lambda)$ is a quasi-periodic function of λ. Suppose now that we let $\{a\}$ denote the integral part of the least residue of a taken modulo q, so that $\{a\}$ is one of the numbers $0, 1, \cdots, q-1$. The function $\{f(c_1\lambda, \cdots, c_p\lambda)\}$ will then yield a doubly infinite sequence of the integers $0, 1, \cdots, q-1$ for λ an integer, which will possess the characteristic recurrence property desired, and will not be of periodic type unless f happens to be very nearly periodic.

Suppose for example that we take $q = 8$ and identify $0, 1, \cdots, 7$ with x, x^{-1}, y, y^{-1}, w_1 or w_1^{-1}, w_2 or w_2^{-1}, w_3 or w_3^{-1}, w_4 or w_4^{-1}. The apparent ambiguity is unimportant since the symbols w_i, w_i^{-1} alternate. Then we should obtain a symbol for a recurrent motion on our surface, corresponding to any function f.

It is in the nature of this proposed method of construction of recurrent non-periodic motions that they possess a certain

p underlying periods, and may very well be of continuous type. The example proposed by Morse was established by him to be of discontinuous type.

The theory of chapter VII *shows that every motion has a certain set of recurrent motions among its closed connected set of* α *and* ω *limit motions.*

The method there used for constructing such a recurrent motion has its symbolic counterpart, whereby at least one normal symbol of recurrent type can be obtained from either end of any normal symbol.

The question arises in how far the closed connected α and ω sets may be assigned at will. Now it is apparent that since the given motion approaches the ω limit motions asymptotically as time increases, it is possible for a sequence of indefinitely long arcs

$$AB,\ B'C,\ C'D,\ \cdots$$

of the limit motions to be found such that B' is very near to B, C' to C, etc., and such that these arcs taken together trace out arbitrarily closely all of the set of limit motions. To this end we may use the given motion in which each long segment MN, NP, PQ, $\cdots$ is near the ω limit set, so that segments AB, $B'C$, $\cdots$ of this set near MN, NP, $\cdots$ respectively exist, and so near all of the ω limit set. Let us say that any such closed connected set of motions is 'cyclic'.

Such α *and* ω *limit sets are necessarily cyclic. Conversely given any two cyclic sets of motions, there exist in the case at hand everywhere dense motions which have precisely these sets as* α *and* ω *limit sets.*

In fact, construct a symbol such that in one direction it contains a succession of symbols attached to arcs $AB, B'C, C'D, \cdots$ of the ω cyclic set as above, in which these symbols increase in length while at the same time the distances BB', CC', $\cdots$ become smaller and smaller; and do the same in the symbol with respect to the α cyclic set, but operating in the reverse direction. In between the two parts an arbitrary finite symbol may be inserted. Such a symbol evidently corresponds to a motion with the desired property, and these motions are clearly everywhere dense because of the presence of the arbitrary symbol.

Finally there exist non-special motions so that the dynamical problem in hand is transitive.

In order to obtain non-special motions we need only write down normal symbols which contain all allowable sequences of symbols.

The set of non-special motions is of course measurable in the sense of Lebesgue, and it is natural to conjecture that its measure is that of the totality of motions in M, i. e. that the special motions are of measure 0. I have not been able to establish this conjecture.

Thus there is an enormous complexity of types of motion in this geodesic problem on a closed analytic surface of negative curvature; but nevertheless a specific algorithm exists which suffices to describe adequately this complication by means of certain symbols.

Of course the above problem differs from the most interesting class of dynamical problem, typified by the geodesic problem on a convex surface, in that all of the periodic motions are of unstable type. Nevertheless it seems to be typical of the general case in many respects.

12. An integrable case. A well-known integrable problem, discovered by Jacobi, is afforded by the geodesics on a convex ellipsoid.* If this ellipsoid flattens to a limiting elliptical form, a special integrable case of the billiard ball problem (chapter VI, section 6) results. This example is still more concrete inasmuch as the geodesics become broken straight lines with vertices lying on the ellipse and making equal angles with the normal at any vertex.

It is a fact of elementary geometry that if a single segment of such a broken line passes through one focus of the ellipse, alternate segments will continue to pass through alternate

* A very suggestive treatment of certain integrable cases will be found in a paper by Whittaker, *On the Adelphic Integral of the Differential Equations of Dynamics,* Proc. Royal Soc. Edinburgh, vol. 47 (1917).

foci, no matter how far produced. A further well-known fact is that if the system of confocal conics containing the limiting ellipse be drawn, successive segments (or these segments extended) will be tangent to a particular conic of the set, which may be an ellipse or a hyperbola; if these points of tangency lie on an ellipse, the billiard ball will

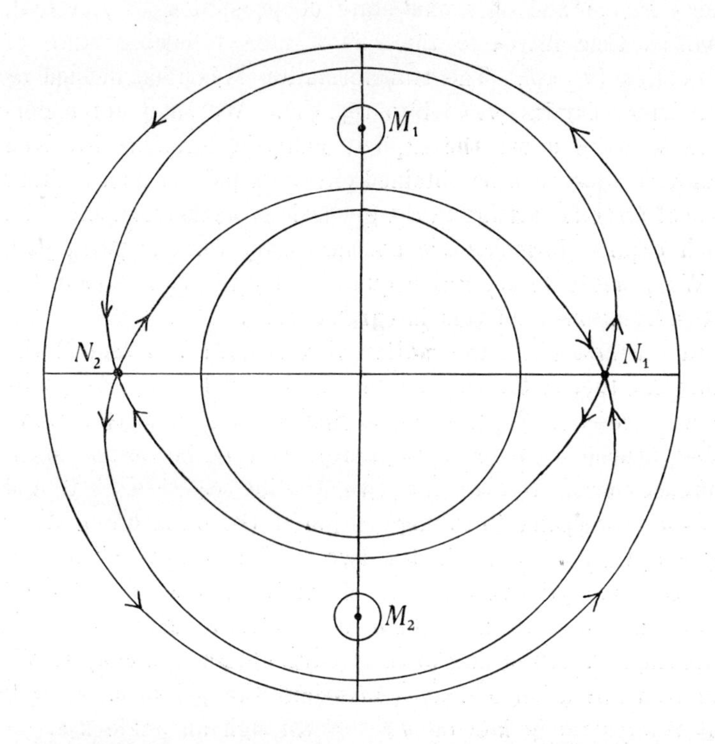

continue to go around the table in one and the sense indefinitely; if these points lie on a hyperbola, the successive points of tangency lie on the two branches in alternation, while the successive segments are between its branches; the major and minor axes constitute two limiting cases of periodic motion.

Suppose now that we make use of the coördinates φ and θ already employed chapter VI, section 7, for setting up the ring transformation T. Here φ denotes an angular coördinate of period 2π which measures arc length along the ellipse while θ measures the angle of projection which the rebounding billiard ball makes with the positive direction of the tangent at the given point of projection on the ellipse.

For any (θ, φ) where $\theta + \pi$, φ may be taken as essentially polar coördinates on the ring-shaped surface of section (figure), there is one and only one state of projection of the ball, and, as time increases there is a next following state of projection (θ_1, φ_1). The transformation T is then defined as that which carries (θ, φ) into (θ_1, φ_1). We shall not undertake to write down the explicit analytic formulas involved although these can be obtained either directly or as a limiting case of formulas arising in the geodesic problem on an ellipsoid. Such explicit formulas are not necessary for our purpose.

We propose to determine qualitatively the character of the transformation T in this integrable case.

In the first place the motion of a billiard ball around the table in either of the two possible senses evidently corresponds to a succession of points on a single closed analytic curve lying near $\theta = 0$ or $\theta = \pi$ respectively, according as φ increases or decreases; the two limiting cases $\theta = 0$ and $\theta = \pi$ correspond to rolling motion of the billiard ball along the elliptical boundary in the two possible senses. Thus we get two analytic families of closed curves which abut on $\theta = 0$ and $\theta = \pi$, and which are invariant under the transformation T. According to the results obtained in chapter VI, the transformation T leaves invariant the points on $\theta = 0$ but rotates the points on $\theta = \pi$ through an angle 2π.

Secondly, if we consider a state of projection which is associated with tangency to a hyperbola, there is one and only one such point of tangency lying on the straight line formed by the segment described by the billiard ball, or that segment extended, and the point of tangency may be continuously varied over the complete hyperbola. There are two kinds of projection, however, corresponding to the values of φ on either of the two elliptical arcs of the given ellipse which lie between the two branches of the hyperbola under consideration. Here each hyperbola gives two closed analytic curves within the ring, and, as the hyperbolas tend to approach the minor axis of the ellipse, we obtain two analytic families of curves closing down upon two points $M_1 = (\pi/2, \pi/2)$, $M_2 = (\pi/2, 3\pi/2)$, corresponding to motion along the minor axis.

Evidently the limiting case of any of these four types of motion is that referred to at the outset when the straight line segments pass through the foci. But the states of projection through either focus correspond to a single closed analytic curve, and these two curves have two points in common, namely the points $N_1 = (\pi/2, 0)$, $N_2 = (\pi/2, \pi)$, corresponding to projection along the major axis.

In this way the points of the ring which represent each state of projection are seen to be divided up as represented in the figure above.

The transformation T leaves the points of the inner boundary invariant and rotates the analytic curves which abut upon it through an angle which increases with distance from the boundary, inasmuch as if θ is increased while φ is held fast, it appears that φ_1 is thereby increased. For the limiting curve of this family made up of two analytic arcs which end at N_1 and N_2, it appears that the transformation T rotates N_1 into N_2 and N_2 into N_1 in a positive sense, while interchanging the two arcs through N_1, N_2.

Similarly the transformation T advances the points of the outer boundary by an angle 2π and rotates the analytic curves which abut upon it by an angle less than 2π, which decreases as the distance from this boundary increases. The limiting curve of this family is made up of two analytic arcs which end at N_1 and N_2, and the transformation T rotates N_1 into N_2, and N_2 into N_1, and interchanges the two arcs.

Examination of the motions which pass through the foci shows that these tend asymptotically toward the major axis in either sense. This agrees with the fact that all points within the inner arcs $N_1 N_2$ are advanced by an angle less than π while those on the outer arcs $N_1 N_2$ are advanced by an angle more than π.

It is clear that there are no invariant points under T.

Let us consider now the various types of motion, and to begin with, one of the type corresponding to a curve of the analytic family abutting upon $\varphi = 0$.

Let us adopt analytic variables (φ, ψ) where ψ varies with the curve of the family. The transformation T takes the form

$$\varphi_1 = F(\varphi, \psi), \qquad \psi_1 = \psi,$$

with Jacobian determinant

$$J = \partial F / \partial \varphi > 0.$$

If the invariant area integral is

$$\int\int I \, d\varphi \, d\psi$$

in these variables (see section 1), we will have

$$I(\varphi_1, \psi_1) \, \partial F / \partial \varphi = I(\varphi, \psi),$$

so that the integral of $\int I \, d\varphi$ over any arc $\psi =$ const., and over the transformed arc has the same value.

Now write

$$\varkappa / 2\pi = \int_0^{\varphi} I(\varphi, \psi) \, d\varphi \Big/ \int_0^{2\pi} I(\varphi, \psi) \, d\varphi,$$

thereby introducing a new analytic angular variable $\varkappa$ of period 2π which can replace φ. The transformation T will take the form

$$\varkappa_1 = \varkappa + \alpha(\psi), \qquad \psi_1 = \psi$$

in these special variables. Here $\alpha(\psi)$ is an analytic function of ψ.

Hence the transformation of each invariant curve of the analytic family which abuts on $\theta = 0$ is essentially a rotation of that curve into itself through an angle α which varies analytically with the curve and increases from 0 along $\theta = 0$ toward a limiting value π. But this variable ψ is not to be regarded as defined along the limiting non-analytic curve of course.

Consequently, if $\alpha(\psi)$ is commensurable with 2π, say $\alpha = 2\pi p/q$, every point of the invariant curve corresponds to a polygonal periodic motion of the billiard ball going p times around the ellipse and having q vertices. Here p is taken relatively prime to q, and p, q take all values for which $p < q/2$.

If $\alpha(\psi)$ is incommensurable with 2π, the entire curve corresponds to a single minimal set of recurrent motions of continuous type.

Furthermore, along the analytic family of invariant curves which abut on $\theta = \pi$, the transformation T is essentially a rotation with rotation number β say, which varies analytically from one invariant curve to another, and which diminishes from 2π at the boundary towards a limiting value π. In this case we have a similar distribution of periodic motions and motions of recurrent type.

If now we turn to the two analytic families of curves which abut on the points M_1 and M_2 respectively, and which are interchanged by T, it is apparent that it is desirable to consider T^2 rather than T, inasmuch as T^2 will leave every curve of either family invariant, and no point on any of these curves can be invariant save under an even iteration of T.

In the same way as before, it follows that the transformation of each of these invariant curves into itself under T^2 is essentially a rotation, of which the rotation number γ varies analytically from curve to curve; at either invariant point M_1, M_2, the value of γ is merely the rotation number for the corresponding stable periodic motion along the minor axis while γ tends toward a limiting value along the limiting non-analytic curve of the family.

If γ is commensurable with 2π, say $\gamma = 2\pi p/q$, every point of the invariant curve corresponds to a polygonal periodic motion going $2q$ times across the ellipse and oscillating in direction p times about the minor axis. If γ

is incommensurable with 2π, the curve corresponds to a single minimal set of recurrent type.

Evidently the points N_1, N_2 correspond to periodic motion along the major axis, and are of unstable type with analytic asymptotic branches represented by the invariant curves through these points. Similarly M_1, M_2 correspond to periodic motion of stable type along the minor axis.

Thus it is seen that the analytic weapons at our disposal have put us in a position to determine the possible types of motion and their interrelations.

Not only so, but the other natural questions which arise can be answered without difficulty.

For example, in the case of motion around the table in either sense is there a unique number which may be properly termed the mean angular rotation? The answer is affirmative in the periodic case, and can also be shown to be so in the non-periodic case. For note that if n denotes the number of vertices passed in any interval of time and if f_n denotes the corresponding increase in the coördinate $\varkappa$ defined earlier then we have clearly

$$\lim_{n=\infty} f_n / n = \alpha$$

where α is the rotation number. Furthermore, if n is large the vertices will be distributed with approximately equal density for $\varkappa$ between 0 and 2π. But the time from any vertex to the next is proportional to the length of the segment and so of the form $l(\varkappa)$, where l is analytic and periodic in $\varkappa$ of period 2π. Hence the total time between the n vertices is

$$l(\varkappa) + l(\varkappa_1) + \cdots + l(\varkappa_{n-1}),$$

which is approximately given by

$$\frac{n}{2\pi} \int_0^{2\pi} l(\varkappa)\, d\varkappa.$$

Thus we conclude that there is a mean angular rotation which has the value

$$\frac{2\pi\,\alpha(\psi)}{\int_0^{2\pi} l(\varkappa)\,d\varkappa}$$

where α and l are definite analytic functions.

13. The concept of integrability. Questions concerning the integrability of a given dynamical problem possess great interest. It is a well known fact that for certain problems, auxiliary analytic relations can be deduced by means of which the solutions of the system of differential equations can be satisfactorily treated, in which case the system may be said to be 'integrable'. When, however, one attempts to formulate a precise definition of integrability, many possibilities appear, each with a certain intrinsic theoretic interest. Let us consider briefly the concept of integrability, not forgetting the dictum of Poincaré, that a system of differential equations is only more or less integrable.

Let us note that in the particular problem just treated there are four periodic motions which play a special role, namely the motions along the two axes of the ellipse and the two motions of rolling around the ellipse.

All other periodic motions fall in analytic families of such motions and so are of highly degenerate type from a formal point of view. But these special motions are isolated and of general type. About these points there will be the usual formal series developments for the coördinates*, and these expressions may of course be taken to converge and to be analytically extended throughout a certain domain of the motion; in fact these properties are merely the counterpart of similar properties of the integrable transformation T, according to which it rotates certain invariant curves surrounding the invariant points in a specific manner.

Let us note also that in this problem four suitable neighborhoods of the four fundamental periodic motions yield the entire manifold M; in fact the two families of motions around

* Of course a discrete integer n figures in the formal series instead of the time t.

the ellipses, the family of motions across the ellipse and the single periodic motion along the major axis yield the totality of motions.

These facts suggest the following (not wholly precise) definition of integrability, based upon a local property and and a non-local property:

A given system of analytic differential equations on a closed analytic manifold M will be said to be integrable if there exists a finite set of periodic motions such that the corresponding complete formal series developments may be taken to converge, and to provide a corresponding analytic representation for every possible motion.

Using this definition as a kind of norm, some reflections suggest themselves.

In the first place it is natural to define 'local integrability' in the vicinity of a periodic motion of general stable type, as that in which the formal series may be taken to converge. Hence the motion is stable in the integrable case, and the explicit formulas yield complete information as to the character of nearby motions.

Now it is conceivable that, although these series do not converge, they may represent asymptotically functions continuous together with some or all of their derivatives near the periodic motions, with the aid of which the differential system can be transformed into a normal form like that of chapter III, section 13, in which $M_1, \cdots, M_n$ are functions of $\xi_1 \eta_1, \cdots, \xi_m \eta_m$, continuous together with certain of their derivatives. Here the qualitative behavior is the same as in the case of convergence. Shall the differential system be called 'locally integrable' in this more general case?

Furthermore, the qualitative behavior of the motions near a periodic motion of general unstable type in the case of two degrees of freedom is essentially independent of convergence or divergence. Shall we call every system locally integrable near such a periodic motion of unstable type?

If the differential system involves a parameter μ, we may inquire into that kind of local integrability in which the formal series are required not only to converge but also to be analytic in μ. It was in this sense that Poincaré established the non-existence of further uniform integrals in the problem of three bodies.* But evidently this definition is logically distinct from that presented in the above definition. Even if a system be non-integrable in this sense, it might perhaps be integrable according to our definition for each particular value of μ.

As far as I am aware, the local non-integrability of no single dynamical problem in the sense formulated above has been hitherto established. We shall, however, establish it in the following way for $m = 1$.

Let us suppose, if possible, that every Hamiltonian problem is locally integrable in the vicinity of a generalized equilibrium point of general stable type (see chapter III). The use of the normal variables shows then that the associated transformation T is essentially a rotation through a variable angle.

Along the invariant analytic curves with rational rotation numbers, all the motions will be periodic in the integrable case, with the same period $2k\pi$. It is this fact upon which we base our argument.

Furthermore, it is in the nature of the transformation T that there are no other periodic motions near equilibrium with this same period $2k\pi$, since the rotation constantly increases with the distance, and the invariant family is represented by a curve which meets every radius vector only once.

Now let us write

$$H = H_0 + \mu H_1$$

where H_0 is the given value of the principal function, μ is a small parameter, and H_1 is an analytic function of p_1, q_1 and t, periodic in t of period 2π, which is free of terms up to the fourth degree in p_1, q_1 at $p_1 = q_1 = 0$, but is otherwise arbitrary. It is readily verified that the multiplier λ and the constant l are independent of μ in the modified problem.

* *Méthodes nouvelles de la Mécanique céleste*, vol. 1, chap. 5.

Furthermore we may assume that H_0 is in normal form

$$H_0 = \lambda p_1 q_1 + \frac{1}{2} l p_1^2 q_1^2,$$

since the normal variables are available. Along any motion of the analytic family, the equations of variation have one periodic solution of period $2k\pi$, and only one, as direct computation shows.

Let μ vary from 0. There are two possibilities. Either the curve representing the periodic analytic family for $\mu = 0$ can be continued analytically, in which case there is a nearby curve for $\mu \neq 0$. Or there will only be a finite number of periodic motions of this period for μ small in absolute value.

In the first case there must obviously be a periodic solution of the equations of variation as to μ, which are obtained by adding non-homogeneous terms $-\partial H_1/\partial q_1$, $\partial H_1/\partial p_1$ to the respective right-hand members of the equations of variation referred to above. But, since $\partial H_1/\partial q_1$, $\partial H_1/\partial p_1$ can be taken almost at pleasure, along any particular periodic motion, the ordinary explicit formulas for the variations δp_1, δq_1 show that no periodic solution will exist in general. In fact the functions δp_1, δq_1 can be expressed as integrals linear in these arbitrary functions, augmented by the general solution of the homogeneous system, one part of which is periodic. Thus there are two conditions to fulfill and only one essential constant, and the condition for compatibility demands that a certain integral over a range $2k\pi$ vanishes, in which the integrand involves $\partial H_1/\partial q_1$, $\partial H_1/\partial p_1$ linearly. Evidently this cannot in general be the case.

Consequently for H_1 suitably chosen and then μ taken arbitrarily small, there will only be a finite number of periodic motions with this period.

But by hypothesis the modified system is integrable. By an other much slighter modification of H we can destroy an analytic periodic family much nearer to the position of generalized equilibrium while not introducing further periodic motions of period $2k\pi$.

Continuing in this manner we set up a limiting admissible principal function H, for which λ and l are unmodified, but for which there are no analytic periodic families belonging to rotation numbers as near to that of the generalized equilibrium as desired. This limiting problem cannot therefore be locally integrable in the specified sense.

Since there are only a denumerable set of periods $2k\pi$ $(k = 1, 2, \cdots)$ which enter into consideration, it is readily seen that no periodic analytic families near equilibrium will exist for suitable H.

In a locally integrable Hamiltonian problem near generalized equilibrium of general stable type, $l \neq 0$, there will exist infinitely many nearby analytic families of motions periodic in a multiple of the fundamental period.

In general a Hamiltonian problem near such a periodic motion will be locally non-integrable, and will possess no analytic families of nearby periodic motions.

It would be possible perhaps, and of considerable interest, to use the same method to show that nearby invariant families asymptotic in opposite senses to one and the same periodic motion do not exist in general. This would eliminate the possibility of invariant families belonging to a rational rotation number, and would establish that in general there is either complete instability or zonal instability.

The same method allows us to establish that multiple periodic motions do not exist in general for dynamical problems of this type.

CHAPTER IX

THE PROBLEM OF THREE BODIES

1. Introductory remarks. The problem of three or more bodies is one of the most celebrated in mathematics, and justly so. Nevertheless until recently the interest in it was directed toward the formal side, and in particular toward the formal solution by means of series.

It was Poincaré* who first obtained brilliant qualitative results, especially with reference to the very special limiting 'restricted problem of three bodies' treated first by Hill. As far as the general problem is concerned, the main achievements of Poincaré were the following: (1) he established the existence of various types of periodic motions by the method of analytic continuation; (2) he proved that, by the very structure of the differential equations, complete trigonometric series would be available; and (3) he pointed out the asymptotic validity of these series. All of these results hold for any Hamiltonian system as well as for the problem of three bodies. Unfortunately an accessory parameter μ is present always in his researches, where for $\mu = 0$ the system is of a special integrable type. Thus the difficulties which arise are partly due to the special nature of the integrable limiting case when two of the three bodies are of mass 0, rather than inherent in the problem itself.

It is not too much to say that the recent work of Sundman† is one of the most remarkable contributions to the problem of three bodies which has ever been made. He proves that, at least if the angular momentum of the bodies is not 0 about every axis through the center of gravity,

* See his *Méthodes nouvelles de la Mécanique céleste.*

† See his *Mémoire sur le problème des trois corps,* Acta Mathematica, vol. 36 (1912); in this connection see also J. Chazy, *Sur l'allure du mouvement dans le problème des trois corps,* Ann. Scient. de l'Ecole Normale Sup. (1922).

the least of the three mutual distances will always exceed a specifiable constant depending on the initial configuration; thus triple collision is proved to be impossible, while it is shown that the singularity at double collision is of removable type. In this way a conjecture of Weierstrass as to the impossibility of triple collision is established, and convergent series valid for all the motion are found for the coördinates and the time. By obtaining such series Sundman 'solved' the problem of three bodies in the sense specified by Painlevé.* As a matter of fact, however, the existence of such series is merely a reflection of the physical fact that triple collision can not occur, and signifies nothing else as to the qualitative nature of the solution.

In the present chapter I propose to take up the problem of three or more bodies, and to endeavor to apply as far as possible the points of view developed in the earlier chapters, and in particular to show what seems to be the real significance of Sundman's results.†

2. The equations of motion and the classical integrals. Let us suppose the three bodies under consideration (taken as particles) to be at the points P_0, P_1, P_2 in space, and to have masses m_0, m_1, m_2 respectively. We denote the distance $P_0 P_1$ by r_2, $P_0 P_2$ by r_1 and $P_1 P_2$ by r_0. If we write

$$U = \frac{m_0 m_1}{r_2} + \frac{m_0 m_2}{r_1} + \frac{m_1 m_2}{r_0}, \tag{1}$$

and if we let x_i, y_i, z_i $(i = 0, 1, 2)$ be the rectangular coördinates of the corresponding bodies P_i, while x_i', y_i', z_i' stand for the components of velocity, the equations of motion may be written as 9 equations of the second order

$$m_i \frac{d^2 x_i}{d t^2} = \frac{\partial U}{\partial x_i}, \quad m_i \frac{d^2 y_i}{d t^2} = \frac{\partial U}{\partial y_i}, \quad m_i \frac{d^2 z_i}{d t^2} = \frac{\partial U}{\partial z_i} \qquad (i = 0, 1, 2), \tag{2}$$

* See his *Leçons sur la théorie analytique des équations différentielles.*

† Most of the new results found in this chapter were announced by me at the Chicago Colloquium in 1920.

which are evidently of Lagrangian form; or as 18 equations of the first order

$$(3)\quad \begin{cases} \dfrac{dx_i}{dt} = x_i', \quad \dfrac{dy_i}{dt} = y_i', \quad \dfrac{dz_i}{dt} = z_i' \quad (i = 0, 1, 2) \\ m_i \dfrac{dx_i'}{dt} = \dfrac{\partial U}{\partial x_i}, \quad m_i \dfrac{dy_i'}{dt} = \dfrac{\partial U}{\partial y_i}, \quad m_i \dfrac{dz_i'}{dt} = \dfrac{\partial U}{\partial z_i} \quad (i = 0, 1, 2), \end{cases}$$

which are of course easily converted by slight modification into Hamiltonian form. We shall not effect this modification, which may be done in the usual way, nor shall we state the usual principles of variation applicable to this case (see chapter II).

The integral expressing the conservation of energy is seen to be

$$(4)\qquad \frac{1}{2}\sum m_i (x_i'^2 + y_i'^2 + z_i'^2) = U - K$$

where K is a constant of integration.

Besides this integral there are of course the 6 integrals of linear momentum expressing the fact that the center of gravity moves with uniform velocity in a straight line; if we take a reference system in which the center of gravity is fixed and at the origin, these integrals reduce to

$$(5)\qquad \begin{aligned} \sum m_i x_i &= \sum m_i y_i = \sum m_i z_i = 0, \\ \sum m_i x_i' &= \sum m_i y_i' = \sum m_i z_i' = 0. \end{aligned}$$

There are also 3 integrals which express the constancy of the total angular momentum about any axis fixed in space. If we take the axes as the coördinate axes, these integrals become

$$(6)\qquad \begin{gathered} \sum m_i (y_i z_i' - z_i y_i') = a, \quad \sum m_i (z_i x_i' - x_i z_i') = b, \\ \sum m_i (x_i y_i' - y_i x_i') = c, \end{gathered}$$

where a, b, c are constants of integration.

These 10 integrals are all the essentially independent integrals which are known.

3. Reduction to the 12th order. The reduction of the system of differential equations (3) to the 12th order may be at once accomplished by use of the integrals of linear momentum as, for instance, by the following method due to Lagrange. Let the coördinates of P_1 with reference to P_0 be (x, y, z) and let the coördinates of P_2 with reference to the center of gravity of P_0 and P_1 be (ξ, η, ζ). If we write for convenience

$$(7)\qquad \begin{gathered} p=\frac{m_1}{m_0+m_1}, \quad q=\frac{m_0}{m_0+m_1}, \\ M=m_0+m_1+m_2, \quad m=\frac{m_0\, m_1}{m_0+m_1}, \quad \mu=\frac{(m_0+m_1)\, m_2}{m_0+m_1+m_2}, \end{gathered}$$

we obtain the explicit formulas of transformation

$$(8)\qquad \begin{gathered} x=x_1-x_0, \quad y=y_1-y_0, \quad z=z_1-z_0, \\ \xi=x_2-p x_1-q x_0, \quad \eta=y_2-p y_1-q y_0, \\ \zeta=z_2-p z_1-q z_0, \end{gathered}$$

together with the inverse formulas,

$$(9)\qquad \left\{\begin{aligned} x_0&=-\frac{m_2}{M}\xi-p x, \quad y_0=-\frac{m_2}{M}\eta-p y, \\ &\qquad z_0=-\frac{m_2}{M}\zeta-p z, \\ x_1&=-\frac{m_2}{M}\xi+q x, \quad y_1=-\frac{m_2}{M}\eta+q y, \\ &\qquad z_1=-\frac{m_2}{M}\zeta+q z, \\ x_2&=\frac{m_0+m_1}{M}\xi, \quad y_2=\frac{m_0+m_1}{M}\eta, \\ &\qquad z_2=\frac{m_0+m_1}{M}\zeta, \end{aligned}\right.$$

which follow with the aid of (5).

The system of the 12th order so obtained may be written in the elegant form

$$
(10)\quad \left\{\begin{array}{lll}
\dfrac{dx}{dt} = x', & \dfrac{dy}{dt} = y', & \dfrac{dz}{dt} = z', \\[2ex]
\dfrac{d\xi}{dt} = \xi', & \dfrac{d\eta}{dt} = \eta', & \dfrac{d\zeta}{dt} = \zeta', \\[2ex]
m\dfrac{dx'}{dt} = \dfrac{\partial U}{\partial x}, & m\dfrac{dy'}{dt} = \dfrac{\partial U}{\partial y}, & m\dfrac{dz'}{dt} = \dfrac{\partial U}{\partial z}, \\[2ex]
\mu\dfrac{d\xi'}{dt} = \dfrac{\partial U}{\partial \xi}, & \mu\dfrac{d\eta'}{dt} = \dfrac{\partial U}{\partial \eta}, & \mu\dfrac{d\zeta'}{dt} = \dfrac{\partial U}{\partial \zeta}.
\end{array}\right.
$$

With these variables the equations (5) may be regarded as satisfied identically while the integrals of angular momentum take the form

$$
(11)\quad \left\{\begin{array}{l}
m(yz' - zy') + \mu(\eta\zeta' - \zeta\eta') = a, \\
m(zx' - xz') + \mu(\zeta\xi' - \xi\zeta') = b, \\
m(xy' - yx') + \mu(\xi\eta' - \eta\xi') = c,
\end{array}\right.
$$

and the integral of energy is

$$
(12)\quad \frac{1}{2}m(x'^2 + y'^2 + z'^2) + \frac{1}{2}\mu(\xi'^2 + \eta'^2 + \zeta'^2) = U - K.
$$

It will be seen that equations (10) may be looked upon as the equations of motion of two particles in space at (x, y, z) and (ξ, η, ζ), with masses m and μ respectively, and in a conservative field of force with potential energy $-U$. These equations can also be derived from either the Lagrangian or Hamiltonian form by use of the variational principles (chapter II).

4. Lagrange's equality. Let us write

$$
(13)\quad R^2 = (m_0 m_1 r_2^2 + m_0 m_2 r_1^2 + m_1 m_2 r_0^2)/M = mr^2 + \mu\varrho^2,
$$

where

$$
(14)\quad r^2 = x^2 + y^2 + z^2, \qquad \varrho^2 = \xi^2 + \eta^2 + \zeta^2.
$$

If now we substitute in (13) the explicit values of r^2 and ϱ^2 obtained from (14), and differentiate twice, there results an equality due to Lagrange,

$$\frac{d^2 R^2}{dt^2} = 2(U - 2K) \tag{15}$$

when use is made of (10) and (12); it is to be observed that U is homogeneous of dimensions -1 in $x, y, z, \xi, \eta, \zeta$ so that

$$x\frac{\partial U}{\partial x} + y\frac{\partial U}{\partial y} + z\frac{\partial U}{\partial z} + \xi\frac{\partial U}{\partial \xi} + \eta\frac{\partial U}{\partial \eta} + \zeta\frac{\partial U}{\partial \zeta} = -U.$$

5. Sundman's inequality. In order to arrive at Sundman's inequality, we propose to seek an upper bound for $(dR/dt)^2$ when $x, y, z, \xi, \eta, \zeta$ are regarded as given quantities while $x', y', z', \xi', \eta', \zeta'$ are to vary at pleasure except that they are to yield the given values of the constant K of energy and of the constants a, b, c of angular momentum. This is a purely algebraic problem.

We have

$$RR' = mrr' + \mu\varrho\varrho',$$

whence

$$R^2 R'^2 = (mr^2 + \mu\varrho^2)(mr'^2 + \mu\varrho'^2) - m\mu(r\varrho' - \varrho r')^2,$$

which may be written

$$R'^2 = mr'^2 + \mu\varrho'^2 - \frac{m\mu}{R^2}(r\varrho' - \varrho r')^2.$$

Furthermore we have the obvious identities

$$\begin{aligned} x'^2 + y'^2 + z'^2 \\ &= r'^2 + \frac{1}{r^2}[(yz' - zy')^2 + (zx' - xz')^2 + (xy' - yx')^2], \end{aligned}$$

$$\begin{aligned} \xi'^2 + \eta'^2 + \zeta'^2 \\ &= \varrho'^2 + \frac{1}{\varrho^2}[(\eta\zeta' - \zeta\eta')^2 + (\zeta\xi' - \xi\zeta')^2 + (\xi\eta' - \eta\xi')^2]. \end{aligned}$$

Multiplying these last two equations through by m and μ respectively, and subtracting them, member for member, from the preceding equation, there results the equation

$$R'^2 + P = 2(U - K) \tag{16}$$

where P (to be minimized) is a sum of seven squares,

$$\begin{aligned} P = {} & \frac{m}{r^2}[(yz' - zy')^2 + (zx' - xz')^2 + (xy' - yx')^2] \\ & + \frac{\mu}{\varrho^2}[(\eta\zeta' - \zeta\eta')^2 + (\zeta\xi' - \xi\zeta')^2 + (\xi\eta' - \eta\xi')^2] \\ & + \frac{m\mu}{R^2}(r\varrho' - \varrho r')^2. \end{aligned} \tag{17}$$

Here the energy integral (12) has been made use of.

From this relation due to Sundman we may derive the inequality which plays a fundamental part in his work and in the present chapter.

If we write

$$U = yz' - zy', \qquad V = \eta\zeta' - \zeta\eta',$$

it will be observed that there are two terms in P of the form

$$S = \frac{m}{r^2}U^2 + \frac{\mu}{\varrho^2}V^2,$$

while the first integral of angular momentum yields

$$mU + \mu V = a.$$

It is easily found that the minimum value of S when U and V vary subject to the restriction just written, while r and ϱ remain fixed, is a^2/R^2. Similarly there are two other analogous pairs of terms with minimum values b^2/R^2, c^2/R^2 respectively. Hence we conclude that we have

$$P \geqq f^2/R^2, \tag{18}$$

$$f^2 = a^2 + b^2 + c^2. \tag{19}$$

Suppose now that we eliminate U between Sundman's equality (16) and Lagrange's equality (15). This gives us

$$2RR'' + R'^2 + 2K = P,$$

whence, by using (18), we obtain the inequality referred to:

$$2RR''+R'^2+2K \geqq \frac{f^2}{R^2}. \tag{20}$$

If we define the auxiliary function of Sundman,

$$H = RR'^2+2KR+\frac{f^2}{R}, \tag{21}$$

the inequality (20) enables us to infer the relation

$$H' = FR' \qquad (F \geqq 0). \tag{22}$$

Hence H increases (or at least does not decrease) as R increases, and decreases (or at least does not increase) as R decreases. This is the consequence which is of fundamental inportance in what follows.

6. The possibility of collision. Thus far we have been taking for granted the existence of solutions in the ordinary sense. In fact, inspection of the differential equations shows the existence of a unique analytic solution for which the coördinates and velocities have assigned values at $t = t_0$, provided that the bodies P_0, P_1, P_2, are geometrically distinct. In the case of the coincidence of two or three of these bodies, the right-hand members of the differential equations are no longer analytic, or even defined, so that the existence theorems of chapter I fail to apply.

But, according to the results there obtained, either these solutions can be continued for all values of the time, or (for example), as t increases, continuation is only possible up to $\bar{t}$.

Let us consider this possibility in the light of the elementary existence theorems.

In the 18-dimensional manifold of states of motion associated with the 18 dependent variables

$$x_i,\ y_i,\ z_i,\quad x_i',\ y_i',\ z_i' \qquad (i = 0, 1, 2),$$

we need to exclude the three 15-dimensional analytic manifolds

$$r_i = 0 \qquad (i = 0, 1, 2).$$

The remaining region is open towards infinity and along these excluded boundary manifolds.

According to the results obtained, indefinite analytic extension of a particular motion will be possible unless as t approaches a certain critical value $\bar{t}$, the corresponding point P approaches the boundary of the open region specified.

Now suppose if possible that the least of the three mutual distances does not approach 0 as t approaches $\bar{t}$; here it is not implied that a specific mutual distance such as $P_0 P_1$ remains least near to $\bar{t}$. We can then find positions of the three bodies for t arbitrarily near to $\bar{t}$, for which the three mutual distances exceed a definite positive constant d. But by the energy integral relation (4), in which

$$U < (m_0 m_1 + m_0 m_2 + m_1 m_2)/d,$$

it is clear that the velocities x_i', y_i', z_i' are uniformly limited. It is physically obvious that for such an initial condition, continuation of the motion is possible for an interval of time independent of the particular mutual distances or velocities, because of the character of the forces which enter; we shall not stop to obtain an explicit expression for such an interval on the basis of our first existence theorem. Thus a contradiction results.

Analytic continuation of a particular motion in the problem of three bodies will be possible unless as t *approaches a certain value* $\bar{t}$, *the least of the three mutual distances approaches* 0.

At this stage it is desirable to revert to Lagrange's equality (15). As t approaches $\bar{t}$, U becomes positively infinite of course. Hence if we represent R^2 as a function of t in the plane by taking t and R^2 as rectangular coördinates, the corresponding curve will be concave upwards for t sufficiently near $\bar{t}$. Therefore R^2 either becomes infinite, or tends toward a finite positive value, or approaches 0.

The first case is manifestly impossible, since one of the bodies would then recede indefinitely far from the two which approach coincidence as t approaches $\bar{t}$; and such a state of affairs is impossible because of the fact that the forces on the distant body are bounded in magnitude.

In the second case it is clear that a particular distance approaches 0, for instance r_2, while the other two approach definite equal limiting values. This is the case of double collision. Since the forces on the non-colliding body are finite near collision, it approaches a definite limiting position; and thus the other two colliding bodies approach a corresponding limiting position, since the center of gravity may be taken fixed and at the origin in the space of the three bodies.

In the third case we have triple collision of course, and this takes place at the origin. However if the constant f is not 0, triple collision cannot take place, as follows from (22) immediately. For it is seen that dR^2/dt will be negative for t near $\bar{t}$ in the case of triple collision, since d^2R^2/dt^2 is positive by Lagrange's equality (15). Hence H will decrease with R (or at least not increases) as t approaches $\bar{t}$. But inspection of H shows that H becomes positively infinite as R approaches 0. Thus a contradiction is reached.

As t approaches $\bar{t}$, there is either double collision between a definite pair of the bodies at a definite point, while the third body approaches a definite distinct point, or there is triple collision at the common center of gravity. If, however, f is not 0, *i. e., if the angular momentum of the three bodies about every axis in space is not constantly* 0, *triple collision can never take place at $\bar{t}$.*

Henceforth we shall make the assumption $f > 0$, thereby eliminating the possibility of triple collision in the sense above specified.

This assumption may be looked upon as merely confining attention to the general case. In fact it is readily proved that in the case $f = 0$, the motion is essentially in a fixed plane. Thus immediate reduction of the problem is possible. Moreover in the case $f = 0$ the angular momentum about a perpendicular to the plane of motion at the center of gravity vanishes. Thus we are only excluding a special case of motion in a plane. The case excluded is of great interest and should be given thorough consideration on its own account.

7. Indefinite continuation of the motions. In the general case under consideration it is thus plain that any motion can be continued up to a double collision.

We propose now to take up briefly the case of double collision in order to render it physically plausible that the motion admits of continuation beyond such a double collision in a certain definite manner. Analytic weapons sufficiently powerful to deal with the singularity of double collision were first developed by Sundman (loc. cit.). A different method of attack, not going outside of the domain of equations of usual dynamical type, has since been obtained by Levi-Civita.* A rigorous treatment of the question will not be attempted here, but the analytic details can be supplied without difficulty on the basis of the researches of Sundman or Levi-Civita.

Let us suppose that the bodies P_0 and P_1 collide for instance, while P_2 is at a distance away. The motion of P_0 and P_1 near collision will clearly be essentially as in the two body problem. What we propose to do is to ignore the disturbing forces due to P_2 during the near approach of P_0 and P_1 to collision, i. e. to replace U by its single component $m_0 m_1 / r_2$, and then to take it for granted that the situation is of essentially the same nature in the actual case.

But if the motion of P_0 and P_1 were just as in the two body problem, their center of gravity would move with uniform velocity in a straight line, while, relative to this point, P_0 and P_1 would move in a fixed straight line until they collide. More precisely, P_0 and P_1 will be at distances inversely proportional to their masses from the center of gravity, while their squared relative velocity is $2(m_0 + m_1)/r_2$ increased by a certain constant whose value depends on the total energy relative to the center of gravity. The motion relative to the center of gravity will be thought of as merely

* *Sur la régularization du probléme des trois corps*, Acta Mathematica, vol. 42 (1921).

reversed in direction after collision. In the original reference system the bodies P_0 and P_1 will describe two cusped curves, and will collide at the common cusp; the cuspidal tangents of the two curves are of course oppositely directed, and it would be easy to specify the precise motion near collision by giving the explicit formulas.

Evidently such a motion of collision in the two body problem is completely characterized by the following quantities: (1) the three coördinates of the point of collision; (2) the three velocity components of the center of gravity at collision; (3) the two angular coördinates θ, φ fixing the direction in space of the axis of the cusp described by P_1, which is the same direction as that of the line of motion relative to their center of gravity; (4) the energy constant. Thus 9 coördinates in all are required to characterize uniquely a state of collision in the two body problem. But to specify any state of motion before or after collision it is necessary to give the time τ that has elapsed since collision.

Furthermore, any motion in which the two bodies almost collide can be characterized in a similar way. Here it is supposed that the initial conditions are slightly modified at some time before collision. In the modified motion it is easy to generalize the above coördinates as follows: (1) instead of the coördinates of the point of collision, we may take the coördinates of the center of gravity when the bodies are nearest to one another; (2) the corresponding velocity components of the center of gravity may be used as before; (3) the angular coördinates θ, φ may refer to the direction of the transverse axis of the conics described relative to the center of gravity; (4) the constant of total energy may be used as before. When the motion is modified slightly in this manner, these 9 coördinates will be only slightly modified.

In addition to these 9 coördinates, the plane of the relative motion must be fixed by a further angular coördinate ψ, and the perihelion distance p must be specified. This gives 11 coördinates to fix upon a particular motion of the two bodies in general position. In order to specify a particular state of motion it is sufficient to specify the time τ measured from perihelion passage.

The coördinate p is not available in the special case of circular motion relative to the center of gravity but this possibility does not arise near a state of collision of the type under consideration.

Hence we find 12 appropriate coördinates in all, corresponding of course to the fact that we have a system of differential equations of the 12th order in the two-body problem.

Let us consider the coördinates in the two-body problem somewhat more attentively. The 6 coördinates determining the position of the center of gravity at nearest approach are obviously unrestricted coördinates. In other words, these sets of 6 coördinates are in one-to-one correspondence with the neighborhood of a point in 6-dimensional space. Similarly the 2 coördinates fixing the axial direction are in one-to-one correspondence with the neighborhood of a point of the θ, φ sphere and are thus unrestricted in the same sense; and so are the total energy and the time τ of course. On the other hand, the perihelion distance p is always positive, and as p approaches 0, the motion approaches that of a definite motion of collision, independently of the coördinate ψ which fixes the plane of the motion. Suppose then that we introduce the following coördinates

$$\alpha = p \cos \psi, \qquad \beta = p \sin \psi,$$

as coördinates serving to replace p and ψ. Collision is then characterized by the conditions

$$\alpha = \beta = 0.$$

The new coördinates α, β are, however, unrestricted.

Consequently in the problem of two bodies, the states of motions near a particular state of collision are in one-to-one, continuous correspondence with the neighborhood of a point in a 12-dimensional space. With this representation the states of motions at collision constitute a 9-dimensional surface through the point.

It is obvious that in a certain sense the singularity of collision is removed by the use of the above coördinates.*

Let us return now to the problem of three bodies in the case under consideration when two and only two of the bodies, say P_0 and P_1, collide. For the motion of collision, we must have as before a definite point of collision, a definite vector velocity of their center of gravity at collision, a cuspidal direction in which collision takes place, and finally a limiting total energy. Furthermore any state of motion before or after collision is characterized by the elapsed time τ.

For motions near a motion of collision, these 9 coördinates admit of simple generalization. For example the instant of 'perihelion' passage can be fixed as that at which the distance $P_0 P_1$ is a minimum, and in this way the position and velocity coördinates of the center of gravity, the axial coördinates, and the perihelion distance can be defined at once, and also the energy constant. The angular coördinate ψ can be taken as that given by the plane which bisects the small dihedral angle defined by the two planes through $P_0 P_1$ and the velocity vectors at P_0, P_1 respectively relative to their center of gravity. The time τ is defined as before. The coördinates p, ψ may be replaced by α, β of course.

Thus on the basis of physical reasoning it appears certain that the singularity of double collision is of removable type, and that the states of motion at double collision form three 15-*dimensional (analytic) sub-manifolds in the* 18-*dimensional manifold* M_{18} *of states of motion, corresponding to the collisions of* P_0 *and* P_1, *of* P_0 *and* P_2, *and of* P_1 *and* P_2 *respectively.*

When the manifold of states of motion is augmented by the adjunction of the parts of the boundary corresponding to double collision, it is obvious that indefinite analytic con-

* For actual removal of the singularity by analytic transformation in the two body problem and similar problems, see Levi-Civita, *Traiettorie singolari ed urti nel problema ristretto dei tre corpi*, Annali di Mathematica, ser. 3, vol. 9 (1903).

tinuation of a motion is possible unless, as t approaches a certain value $\bar{t}$ ($t<\bar{t}$ say), there are an infinite number of double collisions. Let us eliminate this possibility for the case $f>0$, which is under consideration.

In the first place we observe that not only R but also R' must be continuous at double collision. In fact the differential equations themselves show that $d^2\xi/dt^2$, $d^2\eta/dt^2$, $d^2\zeta/dt^2$ are continuous at collision so that ϱ' as well as ϱ must be continuous. On the other hand r' will not be; but, since we have

$$r^2 r'^2 = (xx'+yy'+zz')^2 \leqq (x^2+y^2+z^2)(x'^2+y'^2+z'^2)$$
$$\leqq \frac{r^2}{2m}(U+|K|)$$

on account of the energy integral (12), it is clear that rr' is continuous and vanishes at collision. Hence R' is continuous at collision, having the value $\mu\varrho\varrho'/R$, as follows from (13).

Secondly, as t approaches $\bar{t}$, the least r_i must approach 0. Otherwise we should have $r_i>d>0$ ($i=0, 1, 2$) indefinitely near $\bar{t}$. We have already seen that, because of the energy integral, this would require $x', y', z', \xi', \eta', \zeta'$ to be limited, so that continuation of the motion during a definite interval of time, dependent only on d, would be possible without collision. This is absurd.

Thirdly, R must approach a finite limit as t approaches $\bar{t}$, as follows from Lagrange's equality (15), just as in the case of approach to double collision, inasmuch as R' and R are both continuous at double collision. Reasoning on the basis of Sundman's inequality (22) in the same way as before, we infer also that R cannot approach 0 as t approaches $\bar{t}$.

Hence we conclude that as t approaches $\bar{t}$, the body P_2 approaches a definite limiting position distinct from the corresponding definite limiting coincident position of P_0 and P_1. But it is physically obvious, and might readily be established analytically, that there can only be a finite number of collisions for $t<\bar{t}$ in such a case. Thus a contradiction arises.

In the augmented manifold of states of motion M_{18}, indefinite continuation of every motion for which $f > 0$ is possible in either sense of time. In the case $f = 0$, continuation can only be terminated by triple collision.

Hitherto we have dealt with only the 18-dimensional manifold M_{18}. It is easy to modify the above results so as to apply to the manifold M_{12}, obtained when only those motions are considered for which the center of gravity of P_0, P_1, P_2 lies at the origin. In this case the six coördinates fixing the position and velocity of the center of gravity of P_0 and P_1, for instance, determine these coördinates for P_2.

Entirely similar results obtain in the 12-dimensional manifold M_{12} obtained by fixing upon those motions for which the center of gravity of the three bodies lies at the origin.

As remarked earlier, these results can be fully established by use of the explicit regularizations effected by Sundman or Levi-Civita. An inspection of the formulas leads to the following additional conclusion:

In the augmented manifold M_{18} not only are the states of motion at collision to be regarded as constituted by three 15-dimensional analytic manifolds, but the curves of motion are also to be regarded as analytic and as varying analytically with the initial point and interval, provided this interval be measured by such a parameter as u where

$$t = \int r_0 r_1 r_2 \, du.$$

8. Further properties of the motions. The case $K < 0$ is immediately disposed of, so far as the general qualitative character of the motions are concerned. Lagrange's equality (15) insures that $d^2 R^2/dt^2$ will then exceed $4|K|$. Hence R^2, when plotted as a function of t in the t, R^2 plane of rectangular coördinates, yields a curve with a single minimum which is everywhere concave upwards and rises indefinitely.

Evidently the same conclusion holds for $K = 0$, at least unless U approaches 0. But this can only happen if all three mutual distances increase indefinitely.

In the case $K \leqq 0$, $f > 0$, *at least two, if not all three, of the mutual distances increase indefinitely as time increases and decreases. In the case* $K \leqq 0$, $f = 0$, *the same is true unless the motion terminates in triple collision in one direction of the time.*

A fuller qualitative consideration of the motions $K \leqq 0$ is obviously desirable. But on account of the results just stated it seems proper to consider this case as 'solved' in the qualitative sense.

Henceforth we shall confine attention to the case $f > 0$, $K > 0$, i. e. to the case when the angular momentum of the three bodies about every line through the center of gravity is not constantly 0, and the potential energy is insufficient to allow all three mutual distances to increase indefinitely.

The case $f = 0$, $K > 0$ thus remains. Here the motion is essentially in one plane, and it may be possible to obtain results similar to those here obtained in the case $f > 0$, $K > 0$ by suitable refinement of Sundman's inequality.

We proceed to develop some of the simple and important properties of the motion in the case $f > 0$, $K > 0$.

In the case $f > 0$, $K > 0$ *the least of the three mutual distances cannot exceed* $M^2/(3K)$.

The proof is immediate. By the energy integral (12), U is at least as great as K. But r_0, r_1, r_2 are at least as great as r, the least distance. Hence we obtain

$$(m_0 m_1 + m_0 m_2 + m_1 m_2)/r \geqq K.$$

The numerator on the left is not more than $M^2/3$, whence the stated inequality follows at once.

In the case $f > 0$, $K > 0$, *the largest distance* r_i *will necessarily exceed* k *times the smallest distance* r_j, *provided that*

$$R \leqq 2 m^{*1/2} f^2/(k^2 M^3) \quad \text{or} \quad R \geqq k M^{5/2}/(3K),$$

where m^* *denotes the least of the three masses* m_0, m_1, m_2.

To establish this fact, let k_1 denote the actual ratio of the largest to the smallest distance. Then we have at once

$$R^2 \leqq (m_0 m_1 + m_0 m_2 + m_1 m_2) k_1^2 r^2 / M \leqq M k_1^2 r^2 / 3$$

where r denotes the smallest distance. Likewise we find by a similar calculation

$$U \leqq (m_0 m_1 + m_0 m_2 + m_1 m_2) / r \leqq M^2 / 3r.$$

But Sundman's equality (16) together with (18) gives

$$f^2 / R^2 < 2U.$$

If we employ the inequalities for R^2 and U derived above, this gives readily

$$r > 4 f^2 / (k_1^2 M^3).$$

But inasmuch as R is at least $m^{1/2} r$, while m in turn is at least half of the least mass m^* (see (7)), we find

$$R > 2 m^{*1/2} f^2 / (k_1^2 M^3).$$

Consequently if R is at most of the first stated value, we infer at once that k_1 exceeds k. This proves the first of the two results.

In order to prove the second result, let $\bar{r}$ denote the greatest distance. We then obtain

$$R^2 \leqq (m_0 m_1 + m_0 m_2 + m_1 m_2) \bar{r}^2 / M \leqq M \bar{r}^2 / 3,$$

whence there results

$$\bar{r} > R / M^{1/2}.$$

If we use the inequality already derived for the least distance r, in combination with the one just written, we find

$$k_1 > 3 K R / M^{5/2}.$$

Hence if R is at least of the second value, k_1 will exceed k. This is the second result to be proved.

In the case $f > 0$, $K > 0$, any part of the curve $R = R(t)$, (t, R, rectangular coördinates) for which $R < f / (2^{1/2} K^{1/2})$ consists *of a finite arc, concave upwards and with a single minimum. If $R = R_0$ gives this minimum, the curve rises on either side until*

$$R > f^2/(2KR_0),$$

with corresponding slope R' at least as great as demanded by the inequality

$$R'^2 \geqq \frac{R - R_0}{R}\left[\frac{f^2}{R_0 R} - 2K\right]$$

at every intermediate stage.

To prove this statement, we observe first that when R is restricted as in the first part, R cannot be a constant. In fact if it were, Lagrange's equality (15) would yield $U = 2K$. But the combination of Sundman's equality (16) and of (18) with the equation $U = 2K$ would give

$$f^2/R^2 \leqq 2K,$$

in contradiction with the limitation imposed upon R. The same kind of argument shows that if R' vanishes when R is so restricted, then R'' must be positive. For otherwise, by using Lagrange's equality, we find $U < 2K$, and thence by using Sundman's equality (16) and (18) we are led to the contradictory conclusion written above.

If there is a point $R' = 0$ along the arc under consideration, it corresponds to a proper minimum. On either side of it H (section 5) will increase (or at least not decrease) with R, until a second point $R' = 0$ reached for $R = R_1$. Hence we obtain

$$2KR_1 + \frac{f^2}{R_1} > 2KR_0 + \frac{f^2}{R_0}$$

whence, since $R_1 > R_0$,

$$2K > \frac{f^2}{R_0 R_1}.$$

In this case R does increase until the specified value is passed. Furthermore until this happens, H is as great as H_0.

This fact demonstrates that R'^2 is as great at every stage as stated, so that R must finally so increase.

The case when $R' \neq 0$ anywhere along the arc can be eliminated. Here H must decrease (or at least not increase) with decreasing R. Consequently R cannot approach 0, since H then becomes infinite. As R approaches its lower limit R_0, R' will approach 0. Consequently we infer that the inequality of the statement for R'^2 continues to hold if R_0 be defined in this manner.

But this kind of asymptotic approach to $R = R_0$ as t increases (or decreases) indefinitely is impossible. This impossibility may be made evident as follows. In the inequality $H \geqq H_0$ we may replace the inequality sign by the equality sign. Thereby we define a new curve $R = R(t)$ whose slope for any R is not greater in numerical value than that along the actual curve under consideration. Hence the new curve so defined approaches the t axis less rapidly, and must also approach $R = R_0$ asymptotically as follows from the equation $H = H_0$. But, by differentiation of this equation as to t, there results

$$2RR'' + R'^2 + 2K - \frac{f^2}{R^2} = 0.$$

Hence as t approaches infinity, and R, R' approach R_0, 0, it is clear that R'' would approach a definite positive quantity, which is absurd.

The results thus far obtained may be regarded as concerned with motions in which the three bodies are all near together at some instant $t = t_0$, the amount of separation being measured by R. The bodies will separate in such a way that R increases, and very rapidly as long as R is not too large or small, until R has become very large.

We turn next to derive somewhat analogous results when at least one of the three mutual distances is large. Here it is convenient to use the quantity ϱ instead of R, but it is to be borne in mind that r denotes the smallest of the three distances in what follows.

In the case $f > 0$, $K > 0$ *as long as* $\varrho \geqq 2M^2/(3K)$, *one and the same distance* r_i *is the least distance.*

Under this condition it follows that ϱ is at least twice the least of the distances $r = r_2$. Hence r_0 and r_1 exceed r, since ϱ is the distance from P_2 to the center of gravity of P_0 and P_1. But when r_0 and r_1 are greater than r_2, one and the same distance r_2 remains least.

In the case $f > 0$, $K > 0$, *for* $\varrho \geqq 2M^2/(3K)$, *the inequality*

$$\varrho'' > -8M/\varrho^2$$

obtains. If for any such value of ϱ, *we have*

$$\varrho' \geqq 4M^{1/2}/\varrho^{1/2},$$

ϱ *will constantly increase without bound.*

We begin with the identity

$$\varrho\varrho'' + \varrho'^2 = \xi\xi'' + \eta\eta'' + \zeta\zeta'' + \xi'^2 + \eta'^2 + \zeta'^2.$$

The last three terms on the right give the square of the velocity of the point (ξ, η, ζ), while ϱ'^2 is the square of the radial velocity and is therefore not greater. By virtue of this fact and the differential equations (10) we obtain

$$\varrho\varrho'' \geqq \frac{1}{\mu}\left(\xi\frac{\partial U}{\partial \xi} + \eta\frac{\partial U}{\partial \eta} + \zeta\frac{\partial U}{\partial \zeta}\right).$$

But the terms in parenthesis on the right are precisely $\varrho\,\partial U/\partial n$ where P_2 is taken to vary by a distance n along the straight line which joins P_2 to the center of gravity of P_0 and P_1. Clearly the rate of change of r_0 and r_1 with respect to n cannot exceed 1 in absolute value, and we infer

$$\varrho\varrho'' \geqq -\frac{\varrho}{\mu}\left(\frac{m_1 m_2}{r_0^2} + \frac{m_0 m_2}{r_1^2}\right) > -M\varrho\left(\frac{1}{r_0^2} + \frac{1}{r_1^2}\right)$$

(see (7)). Now in the case under consideration r_0 and r_1 exceed $\varrho - r$ and therefore $\varrho/2$. This leads to the first inequality to be proved.

Instead of continuing analytically we need simply observe that this inequality may be looked upon as requiring that a particle moves along a ϱ axis acted upon by a force towards the origin which does not exceed the gravitational force due to a mass $8M$. But in this case it is obvious that the particle will recede indefinitely provided that the initial velocity outward is as great as the velocity of fall from infinity under the attraction of such a mass. This is precisely the fact stated.

It should be noted that since the initial value of ϱ is as great as $2M^2/(3K)$, ϱ continues greater than this quantity, and accordingly one and the same distance r is the least of the three distances always.

We propose next to combine these results in order to show that, for the minimum R_0 sufficiently small, R and ϱ increase indefinitely. The qualitative basis of the reasoning is obvious. According to what has been proved, for R^* and $R^{*\prime}$ arbitrarily large a positive R_0 can be chosen so small that all motions for which the minimum R is not more than R_0 correspond to an R which increases from the minimum to R^* and has, for $R = R^*$, a derivative R' which is at least as great as $R^{*\prime}$. This means of course that ϱ^* is arbitrarily large since

$$\lim_{R=\infty} R/\varrho = (m_0 m_2 + m_1 m_2)^{1/2}$$

uniformly. Furthermore since the relation

$$R R' = m r r' + \mu \varrho \varrho'$$

obtains, it is clear that $|\varrho \varrho'|$ must be large, and in particular $|\varrho'|$ must be large, provided that $|r r'|$ is uniformly bounded. But we have

$$r'^2 \leqq x'^2 + y'^2 + z'^2 < 2U/m$$

by the energy integral (12). Hence

$$r^2 r'^2 < 2(m_0 m_1 + m_0 m_2 + m_1 m_2)\, r/m < 2M^2 r/m^*$$

since m exceeds one half of the least mass m^*. Thus we find

$$|rr'| < M^2/(K^{1/2} m^{*1/2}),$$

and thereby establish the fact that $|rr'|$ is uniformly bounded.

For $f > 0$, $K > 0$, if R_0 is taken sufficiently small, every motion for which the three bodies approach so closely that $R \leqq R_0$ at some instant is such that two of the distances r_0, r_1 become infinite with t while r_2 remains less than $M^2/(3K)$.

We shall not pause to develop an analytic formula which yields a suitable R_0, although the specific results found above would supply the basis for such a computation.

There is an interesting question to which we wish to refer briefly in conclusion. Which one of the three bodies will recede indefinitely from the other two nearby bodies, in the case of a near approach to triple collision? The answer is to be found in the following statement:

Any motion of the above type is characterized by the property that one and the same body P_2 remains relatively remote from the two nearest bodies P_0, P_1 throughout the entire motion.

The truth of this fact is readily inferred. At the beginning of this section it was shown that, for R greater or less than fixed values, the ratio of the largest to the smallest distance would be arbitrarily large. Hence we need only consider this intermediate range of values of R. But in such a range, if the ratio of the largest to the smallest side did not remain large for R_0 sufficiently small, there would be configurations of the three bodies in which the distances r_i and the ratios r_i/r_j lie between fixed bounds, no matter how small R_0 is chosen. However, the value of U does not exceed an assignable quantity in such configurations, and thus, by the energy integral (12), the same would be true of the velocities x', y', z', ξ', η', ζ'. Finally it is clear that RR' would not exceed an assignable quantity. But we have established that R' becomes arbitrarily large in such a definite range of values of R, so that this conclusion is absurd.

Evidently there is further work to be done in the more precise determination of the motions on the quantitative side, but the facts developed above are sufficient to show that the only possibility of simultaneous near approach of the three bodies for given $f > 0$, $K > 0$, is that in which the three bodies act as a pair of bodies, one member of which corresponds to a close double pair P_0, P_1, while the second is P_2. The motions of P_2 and the center of gravity of P_0, P_1 are then along nearly hyperbolic paths, while P_0, P_1 move in nearly elliptic paths relative to their center of gravity.

9. On a result of Sundman. Sundman established (loc. cit.) that for given initial coördinates and velocities with $f > 0$, $K > 0$, the quantity $R(t)$ for the corresponding motion will always exceed a specifiable positive constant. This fact is at once evident from the analysis of section 8. In the contrary case we should have indefinitely near approach to triple collision, and thus a motion for which R' is arbitrarily large for the given initial value of R, which is of course absurd.

10. The reduced manifold M_7 of states of motion. Let us turn next to the consideration of the problem of three bodies after use has been made of the 10 known integrals to reduce the system of differential equations from the 18th to the 8th order. In other words the 10 corresponding constants of integration are given fixed values, and attention is directed towards the ∞^7 motions which correspond to the given set of constants. In what follows we shall suppose that not all the constants of angular momentum vanish, and that the constant of energy is positive, i. e. we take $f > 0$, $K > 0$.

The angular momentum vector with components a, b, c will define a spatial direction which plays an important role in the sequel. Evidently two motions which correspond to the same configuration of positions and velocities at some instant, aside from mere angular orientation relative to this axis of angular momentum, will continue to differ merely in this respect. In other words, if φ denotes any angular coördinate which fixes the orientation about the axis of angular momentum, while $u_1, \cdots, u_7$ are any set of relative coördinates which do not involve φ, the differential equations defining the ∞^7 motions take the form

$$du_i/dt = U_i(u_1, \cdots, u_7) \quad (i = 1, \cdots, 7),$$
$$d\varphi/dt = \Phi(u_1, \cdots, u_7).$$

The first set of equations constitutes a system of the 7th order in the coördinates $u_1, \cdots, u_7$, while the last equation enables one to determine φ by a further integration. If it be desired, the time t can be eliminated, and the system becomes of the 6th order,

$$du_i/du_1 = U_i/U_1 \quad (i = 2, 3, \cdots, 7).$$

Thus from the purely formal standpoint the system of the 18th order can be 'reduced' to one of the 6th order.

From the point of view which we shall adopt, there is no essential gain in actually carrying through such a reduction which can be accomplished without affecting the Hamiltonian form.*

Let us consider the augmented manifold M_{18} of states of motion, in which the singularities corresponding to double collision have been removed by the method indicated in section 7.

The boundary of M_{18} is to be regarded as made up of states of motion specified by one of the following possibilities: one of the coördinates x_i, y_i, z_i increases indefinitely in absolute value; the quantity R approaches 0; the energy constant of some pair P_i, P_j of the bodies relative to their center of gravity at the instant increases indefinitely in absolute value. It is clear that points away from the boundary in the specific sense of these three possibilities will have limited coördinates, with not all three distances small; the condition of energy imposed insures that the energy constant relative to the center of gravity of all three bodies is not large in absolute

* See, for instance, Whittaker, *Analytical Dynamics,* chap. 13.

value, while the fact that the relative energy constants are not large means that the nearest pair of bodies must shortly separate to a considerable distance. Thus either all coördinates and velocity components are limited, and none of the mutual distances are small, or else the motion is near such a state in time, and therefore not near to the boundary of M_{18}.

In M_{18} the totality of motions is represented as a steady fluid motion, in which the stream lines correspond to the possible types of motion. When the 10 constants of integration are specified, we are directing attention to the corresponding fluid motion of the sub-manifold M_8 into itself in which the stream lines represent the ∞^7 motions under consideration.

Motions which differ merely in orientation with respect to the axis of angular momentum yield a closed one parameter family of stream lines, corresponding states of which give closed curves; in other words $u_1, \cdots, u_7$ are the same along such a curve, while φ varies from 0 to 2π. In the special case of the Lagrangian equilateral triangle and straight line solutions when the mutual distances are inalterable,* the corresponding closed curve is itself a stream line.

The 'reduced manifold M_7 of states of motion' corresponds to the ∞^7 set of states of motion given by sets of coördinates such as $u_1, \cdots, u_7$, which are distinct except in orientation about the axis of angular momentum.

It is evident that in the original M_{18} the closed curves which give the states of motion differing only in orientation will give ∞^{17} analytic curves, one and only one through each point. Hence if we desire to obtain more precise information as to the possible singularities of M_7, it is only necessary to determine the singularities of M_8. We propose to investigate the singularities of M_8, and thus of M_7, sufficiently to establish the following result:

For general values of $f > 0$, $K > 0$, *the analytic reduced manifold* M_7 *of states of motion is without singularity, and has a boundary specified by the fact that either* R *approaches*

* See Lagrange's paper, *Essai sur le problème des trois corps*, Œuvres, vol. VI.

0 *or* ∞, *or that the energy constant of some pair of the bodies relatively to their center of gravity become indefinitely large and negative.*

Let us first justify briefly the statement made about the boundary of M_7. At some distance from the boundary none of the coördinates can be large since none of the distances r_i are large, and the center of gravity is at the origin. Since the energy constant for the three bodies is given, the partial energy constants cannot be large and positive. Consequently unless one of these partial constants is large and negative, the state of motion is not near the boundary of M_7.

In dealing with the analytic character of M_8, and so of M_7, we can assume that the state of motion under consideration is not a state of double collision. In fact the 'molecule' of states of motion in M_{18} near a state of double collision is carried analytically into a molecule about a modified position, not corresponding to a state of double collision. The invariant sub-manifold M_7, will thus be analytic all along a particular stream line or nowhere along it.

Let us then employ the coördinates $x, y, z, \xi, \eta, \xi, x', y', z', \xi', \eta', \xi'$ which are available in M_{12}, within which we may take M_8 to lie. The sets of these 12 coördinates which satisfy the remaining angular momentum and energy conditions (11) and (12), furnish uniquely the states of motion of M_8 near to the particular motion of M_8 under consideration. It is evident that in general these 4 equations may be solved analytically for any 4 of the 12 variables; i. e. M_8 analytic at the corresponding point.

We can show, however, that for general values of $f > 0$ and $K > 0$ there can be no singularities whatsoever in M_8. Let us choose coördinate axes so that $x = y = \eta = 0$ at the instant under consideration, i. e. the particle P_1 lies in the z direction from P_0, while the line from P_2 to the center of gravity of P_0 and P_1 lies in the x, z plane. Let us attempt to solve the 4 equations for x', y', z', η' as functions of the other variables. The condition that this be possible will be satisfied if the corresponding Jacobian determinant

$$\begin{vmatrix} 0 & 0 & 0 & \xi \\ 0 & -z & 0 & -\zeta \\ z & 0 & 0 & 0 \\ x' & y' & z' & \eta' \end{vmatrix}$$

does not vanish; here we have removed an obvious factor m from the first three columns, and a factor μ from the last column. Thus M_8 is analytic at this point provided that the inequality

$$-\xi z^2 z' \neq 0$$

holds. But it has been pointed out that z is not 0. Furthermore, we can take $\xi \neq 0$ unless P_2 is on the straight line $P_0 P_1$ constantly. And we can take $z' \neq 0$ unless the distance $P_0 P_1$ (and similarly any other distance $P_i P_j$) is a constant. Hence we infer that either M_8 is analytic along the particular stream line under consideration, or the three bodies lie upon a straight line, or at a constant distance from each other, but not in the same straight line.

In the latter case the bodies P_0, P_1, P_2 are known to lie at the vertices of an equilateral triangle in a plane perpendicular to the angular momentum vector; this triangle rotates at a constant angular velocity about its center of gravity. Furthermore it is known that there is one and only one size of triangle of this kind for an assigned angular velocity. Thus there will be in general no such motion for which f and K have the preassigned values.

Similarly in the first case further examination shows that the distances are inalterable. It is known that there are three solutions for an assigned angular velocity, and thus in general no solution for general values of f and k.

In any case the manifold M_7 can only have a singularity at a point corresponding to an equilateral triangle solution or to a straight line solution at constant mutual distances. These possibilities will only arise when certain analytic relations between f and K are satisfied. It is only as f and K vary through these critical values that the nature of M_7 from the standpoint of analysis situs can change.

The manifold M_7 has fundamental importance for the problem of three bodies, but, so far as I know, it has nowhere been studied even with respect to the elementary question of connectivity. The work of Poincaré refers to the existence of certain periodic motions, i. e. of certain closed stream lines in M_7, obtained by the method of analytic continuation from a limiting integrable case of the problem of three bodies; nearby motions, i. e., stream lines in the torus-shaped neighborhood of such a closed stream line, are also considered in relation to the formal series; but he does not consider M_7 in the large.

In conclusion it may be observed that the states of motion in which the three bodies move constantly in a plane through the center of gravity perpendicular to the angular momentum vector, correspond to an invariant sub-manifold M_5 within M_7, which contains the exceptional singularities when these exist. So far as dimensionality is concerned, this manifold M_5 would be suited to form the complete boundary of a surface of section (chapter V) of properly extended type.

11. Types of motion in M_7. The problem of three bodies is distinguished from the type of non-singular problem which we have considered earlier, in that the manifold of states of motion is not closed. The singularity along the boundary cannot be removed by any exercise of analytic ingenuity. In fact consider a tube of stream lines in M_7 described by a 'molecule' of states of motion near triple collision at $t = 0$. It is clear that the molecule tends toward the boundary of M_7 as t increases, since we have then $\lim R = \infty$ according to the results deduced above (section 8). The half tube so generated is then carried into part of itself, and would have to correspond to an infinite value of the invariant 7-dimensional volume integral. This situation does not arise when the manifold of states of motion is closed and non-singular.

More precisely, the stream lines corresponding to motions of near approach to triple collision not only lie wholly near the boundary of M_7, and approach it as t increases or decreases indefinitely, but they fill out three entirely distinct regions of M_7, since for every such motion there is a particular one of the three bodies which recedes indefinitely from the other two bodies.

The stream lines corresponding to near approach to triple collision thus fill three distinct 7-dimensional continua of M_7, *corresponding to the fact that* P_0, P_1, *or* P_2 *may be the relatively distant body during such a motion. These continua lie near to the boundary of* M_7, *and every stream line in them approaches the boundary in either sense of time.*

Of course these continua are not precisely defined until the degree to which triple collision is approached is precisely specified.

It is natural to believe that in this case of indefinite recession, the two nearby bodies have a definite limiting energy constant, orientation of plane of motion, eccentricity, and a limiting linear and angular momentum with reference to the center of gravity of the three bodies. In any case these motions may properly be regarded as to a large extent 'known'.

The very interesting question now arises: Do the motions for which $\lim R = \infty$ in one or both directions of the time fill M_7 densely or only in part? It is important to understand the nature of the difficulty inherent in this question. By actual computation of the motions, it can doubtless be established whether or not a specific motion belongs to one of these continua or not. Certainly, for $|K|$ small, almost all of M_7 would be filled by these continua in consequence of the results obtained in the case $K \leqq 0$. Nevertheless when there exists a single periodic motion in M_7 of stable type, it will not be possible to determine whether or not nearby motions belong to these continua without solving the fundamental problem of stability in this particular case. We have already alluded to the highly difficult character of the problem of stability (chapter VIII), which arises precisely because in a dynamical problem such as the problem of three bodies, formal stability of the first order insures the satisfaction of all the infinitely many further more delicate conditions for complete formal stability.

The question can, however, be put in a very suggestive form, which in my opinion renders it probable that the motions for which $\lim R = \infty$ for $\lim t = +\infty$ fill up M densely, as do those for which $\lim R = \infty$ for $\lim t = -\infty$; because of the reversibility of the system of differential equations, both conjectures must be either true of false.

The manifold M_7 has already been conceived of as a 7-dimensional fluid in steady motion. This fluid must be thought of as having infinite extent and as incompressible, in consequence of the existence of a 7-dimensional volume invariant integral. The three types of motion with near approach to triple collision correspond to three streams which enter M_7 from the infinite region and leave it there.

What is likely to happen to an arbitrary point of the fluid? It seems to me probable that in general such a point will move about until it is caught up by one of these streams and carried away. It may, however, be anticipated that there will be found certain points which remain at rest or move in closed stream lines, and so are not carried off. In conformity with the results of chapter VII, there must then necessarily exist other stream lines which remain near to the closed stream line as time increases or as time decreases. More generally, there will exist recurrent types of stream lines corresponding to recurrent motions, and various other stream lines which remain in their vicinity as time increases or decreases. The stream lines corresponding to such recurrent motions and nearby motions cannot of course approach the boundary of M_7.

For the determination of the distribution of such periodic motions, recurrent motions, and motions in their vicinity, it obvious that elaborate detailed analysis would be necessary. In conclusion we shall merely effect an obvious classification based on the function $R(t)$:

An arbitrary motion in the problem of three bodies for the case $f > 0$, $K > 0$ *is of one of the following types as* t *increases*:

(1) R *increases toward* $+\infty$, *in which case one body recedes indefinitely from the other two, while the near pair remain within finite distance of one another*;

(2) *R tends toward a value $\bar{R}$ while U approaches $2K$, in which case the limiting motion is of special determinable type as in Lagrange's equilateral triangle solution;*

(3) *$R(t)$ is uniformly bounded as in case (2) but oscillatory. Here the motion is wholly one of finite distances and velocities except possibly for occasional double collisions or approach to such collisions, and there necessarily exist periodic or other recurrent motions among the limit motions;*

(4) *$R(t)$ is oscillatory with upper bound $+\infty$ and a positive lower bound. This is an intermediate case in which the motion is one with finite velocities except near occasional double collision or approach to double but not triple collision, while from time to time one of the three bodies recedes arbitrarily far from the near pair only to approach them again later.*

Similar results obviously hold as t decreases.

The only part of this statement calling for any explanation is that if R approaches $\bar{R}$, U approaches $2K$. But this can be proved to follow from Lagrange's equality (15).

12. Extension to $n > 3$ bodies and more general laws of force. In indicating the possibility of generalizing the above results, both in respect to the number of bodies and the law of force, we shall entirely put to one side the question of collision. It would suffice for our purpose, however, if any kind of continuation after multiple collision were possible in which the constants of linear and angular momentum as well as of energy are the same after as before collision, and if also R', where

$$R^2 = \frac{1}{2M} \sum m_i m_j r_{ij}^2,$$

may be regarded as continuous at collision; here the masses of $P_1, \cdots, P_n$ are $m_1, \cdots, m_n$ respectively, while M is the sum of these masses, and r_{ij} denotes the distance $P_i P_j$.

Let the function U of forces be any function of the mutual distances r_{ij}, of dimensions -1 in these distances. For a function U of this type, the original form of differential equations, of the 10 integrals, and of Lagrange's equality

(15) and of the inequality (20) due to Sundman will subsist, provided that f denotes the total angular momentum of the system about the center of gravity. Our main reasoning above was essentially based upon this analytical framework. Hence we can state the following result:

Let U be any analytic function depending on the mutual distances between n bodies P_i, $(i = 1, \cdots, n)$, with coördinates (x_i, y_i, z_i) and masses m_i respectively; let U be furthermore homogeneous of dimensions -1 in these distances. If the n bodies are sufficiently near together, with assigned positive values of the total angular momentum f and the constant of energy K, at least two of the mutual distances will become very large in either sense of the time.

Further consideration shows that the condition of homogeneity upon U can be lightened to the form of an inequality

$$\sum \left(x_i \frac{\partial U}{\partial x_i} + y_i \frac{\partial U}{\partial y_i} + z_i \frac{\partial U}{\partial z_i} \right) \geqq -dU$$

where $0 < d < 2$, without affecting the argument that at least two of the mutual distances become very large.

In this argument the function H has to be generalized to the form

$$H = R^d \left[R'^2 + \frac{f^2}{(2-d) R^2} + 2K \right].$$

I have not attempted to ascertain conditions under which at least two of the mutual distances become infinite.

INDEX